Stone Walls and Sugar Maples

An Ecology For Northeasterners

C. John Burk and Marjorie Holland

Illustrator, Pamela See

Appalachian
Mountain Club

If you have any corrections or comments on this book, please write to:
Stone Walls and Sugar Maples
Appalachian Mountain Club
5 Joy Street
Boston, MA 02108

ISBN 0-910146-22-5

Contents

Acknowledgments

We acknowledge with gratitude the encouragement and assistance in writing the "Ecology Seminar Series" and this book received from the following members of the staff of the Appalachian Mountain Club: Rick Brunswick, Katryn Gabrielson, Nelson Obus, Arlyn Powell, Robert Saunders, Bill McBee, and Meg Schwarz. Mrs. Kathleen Tobey and the office staff of Clark Science Center, particularly Gertrude Ahearn, Julia Britt, Pearl Duffney, Joice Gare, and Polly Laliberté provided invaluable assistance throughout in the preparation of the manuscripts, as did Dolores Mosakewicz. We also thank Lincoln Brower, David Haskell, Robie Hubley, Donald Mader, John Pinto, Marshall Schalk, Allen Curran, and Bruce Spencer for advice and assistance with the illustrations. Also a special thanks to Nelson Obus who wrote the Noise and Sound chapter. Finally our gratitude for the patience and endurance of our spouses, Lâle Burk and Russell Sackett, throughout this project cannot be overstated.

The cover was designed by Lisa Douglis, cover photo by Arlyn Powell, interior design by Robert Saunders, and indexing by Bill McBee. Composition, printing, and binding were done by Heffernan Press.

C. John Burk
Marjorie Holland
Northampton, Massachusetts
October 26, 1979

Introduction

The impetus for this book was a series of articles we wrote for *Appalachia Bulletin*, the monthly magazine of the Appalachian Mountain Club, over the past several years. Called the "Ecology Seminar Series," its purpose was twofold: to introduce Northeasterners to the principles of the science of ecology as those principles apply to their geographic region, and to apply the principles to some of the universal issues that are now confronting all of us.

The book may be divided into several parts. The first presents an overview of the ecological history of the Northeast. The second outlines some principles of the science of ecology. The third discusses the natural habitats likely to be found in the Northeast. And, the fourth explores ecological problems, both regional and worldwide.

To introduce this series of articles, we have added a discussion of the development of the landscape of the northeastern United States, including its vegetation and wildlife. To do so seems essential for the following reason: most of us have acquired, from a variety of sources, an impression (which may approach the strength of conviction) that from the time of the European settlements of the seventeenth century, human populations, utilizing the growing technology of Western civilization, have wreaked havoc on a natural system which had previously existed in a fine state of balance. The New World, we are frequently told, was a paradise which our forefathers destroyed; we must, if we are to survive in it, find our way back to a state of co-existence with and adjustment to those very natural forces our culture has defied.

We do not deny the extent of society's impact on the countryside or the need for reconsideration and reform of much of our behavior. However, the goal of this reassessment cannot be a return to a former golden period of harmony with nature, but rather must be an adjustment to what has always been a dramatically changing and dynamic environment. Even to begin to perceive how these adjustments might be made, we must understand the nature of the processes which have interacted and now interact to produce the land which we inhabit. To address this problem, we would like to examine the following questions: (1) What was the region like when the first European settlers arrived? (2) What forces had created the landscape these settlers perceived? and (3) What changes from the colonial period to the present have been brought about by our activities?

1.

The Northeastern Landscape

Some historians have suggested, with good reasons, that the colonists arrived on the North American continent during the least auspicious centuries of the last few thousand years. From the seventeenth through the nineteenth centuries, climatic conditions worldwide were harsher than at present and worse than they had been for hundreds of years before. During this "Little Ice Age," mountain glaciers were expanding, summer temperatures averaged 1.5 degrees Centigrade colder than at present, and precipitation in winter was as much as twice as heavy as now.

Some sense of the nature of the vegetation at that time can be determined by examining the writings and records of early travelers, surveyors, and writers and by investigating fossil material and pollen preserved at different levels in bogs. Undoubtedly, most of the land was forested and the trees were the same species found at present in each section of the region. Since the colonial settlement there have been changes in the relative proportions of the various tree species and, of course, in the total amount of forested land.

For our purposes we can consider there to be three basic although intergrading divisions. The first is a deciduous woodland in the southern portions of our geographic area; here oaks are the most abundant and distinctive trees. Inland and northward at lower elevations is a more diverse forest type composed of hardwood species, including beech, sugar maple, yellow birch, and the cone-bearing hemlock. Even farther north and at higher elevations is a coniferous forest of red spruce and balsam fir. This

1

forest occurs at sea level along the coast of Maine above Portland and at increasingly higher elevations south along the Appalachian Mountain chain.

These divisions sort themselves out in many locations according to topography, with oaks, for example, prominent on the warmer, southerly slopes of a given hill and hemlock and hardwoods on its cooler northern side. In addition, pines may be extremely important in the woodlands throughout the region, reaching their greatest abundance on sites where there has been previous disturbance. Pines require much light and hence are intolerant of shade, including shade cast by other pines; their seeds, moreover, germinate poorly in organic soils. As a result, pines tend to be excluded from mature forest types. In part, their persistence throughout the Northeast can be attributed to periodic hurricanes, ice storms, fires, or other "catastrophes" which open previously closed forests, providing abundant light and bare soil for a seedbed. The life history of the pine, which includes a frequent and heavy seed set, wide dispersal by wind of the winged seeds, a rapid growth rate in open areas, and the great height of the mature trees, is closely geared to succeeding under these conditions. One substantial area of southern New England formerly supported a large but scattered population of extremely tall white pines, some exceeding two hundred feet in height and towering above hardwoods which were scarcely half as tall. These giant pines often lived to an age of four hundred years or more. They possessed a greater resistance to ice and heavy snows than the hardwoods beneath, and when they did fall, crashing to the forest floor, they opened sites where their offspring could thrive. White pine is easily killed by fire; on sandy soils and where fires, either natural or set by man, are common, the shorter pitch pine, which has fire-resistant cones, takes its place in the forest community.

The deciduous forest covered much of what is now Connecticut, southern Rhode Island and New York, and southeastern and central Massachusetts. White, black, red, and scarlet oak were the dominant trees, in association with American chestnut, hickory, tulip poplar, and on drier sites, the chestnut oak. Red maple, elm, and pin oak were common trees in swampy areas. These forests were greatly affected by the activities of Indians; their open, park-like nature was noted by the Pilgrims and many others and has been a source of speculation for ecologists in the present

2

century. Some believe that the Indians may have set fire deliberately to the woods to make traveling easier and to create a more favorable habitat for game species, including the heath hen, wild turkey, and deer. Others dislike the fire hypothesis, believing that the so-called sprout hardwoods are relict in nature — that is, left behind from a warmer, drier period in the past. One of the most articulate proponents of the latter school, Hugh M. Raup, has commented that ". . . to picture such a wholesale conflagration . . . . as would involve most of the inflammable woods every year, or even every 10 or 20 years, is inconceivable." Even he admits, however, that fire and the activities of the Indians played at least some role in the forests of the area. In addition to at least occasionally burning the forests, they cut wood for fuel, cleared land for village sites and agriculture, and planted or at least encouraged the growth of plants, including the Canada plum, nut-bearing chestnuts, hickories and walnuts, and the Kentucky coffee tree, all of which were useful for food or medicine.

A pine stand in Pennsylvania. / Courtesy of the Arnold Arboretum

The behavior of animals — browsing by deer; feeding, roosting, and nesting activities of wild turkeys and passenger pigeons; and cutting of trees and flooding associated with the work of beavers — must also have helped to form the landscape.

Fires set by man were less important in the mixed hardwoods and hemlock forests, where beech and maple were said to be too wet to burn. In the coniferous forests, which the Indians did not occupy, there was little human influence on the vegetation. These spruce-fir forests are a southerly extension of a broad band of coniferous forest which sweeps across the northern part of the continent. To understand how they and the other vegetation types we have mentioned arrived at their present locations, we must look farther back in time to the end of the last glaciation, when much of New England lay beneath a sheet of ice.

2.

Back to the Ice Age

An observant traveler, driving on a summer day from the summit of Mount Washington down to Boston and then proceeding south along the coast, is able to see in a journey of less than a week a procession of vegetation types which, had the traveler remained fixed at a single vantage point — say Mount Tom in western Massachusetts — would have passed him moving northward through the last 12,000 years or more. Coming down off Mount Washington, our traveler would pass through the alpine zone and then a forest of spruce, fir, and birch, before reaching mixed hardwoods and coniferous trees. Spruce and fir would drop out completely from the forest, and the northern hardwoods and hemlocks would give way in time to a woodland mostly of oak and pine. As the traveler continued south, behind the trappings of civilization — cities, suburban developments, highways, industrial parks — he would glimpse in woodlots and stands of second growth a transition from the northern oak species to more typically southern forms — blackjack and turkey oak on sandy soils, Spanish red oak, water oak and willow oak on the lower, wetter sites. The pines would change, too, white pine giving way to pitch pine and then to loblolly, longleaf, yellow, and other species of southern pine.

Had our traveler stayed as close to the edge of the sea as possible, had he driven from Mount Washington across to the coast of Maine and then proceeded south, he would have found the rocky shoreline replaced by a region of reworked glacial deposits — old moraines and outwash plains with erratic boulders strewn across cleared fields — from Massachusetts to Long Island Sound. From New Jersey on, he would have encountered wide, sandy beaches which separate the Atlantic Ocean from a broad, flat coastal plain.

The way this landscape evolved has been a subject of investigation for more than a century, a magnificent if baffling puzzle to which geologists, biologists, archeologists, geographers, and historians have contributed clues, and over which they argue, sometimes fiercely, in trying to determine how the all-too-scattered bits of evidence mesh. New discoveries and the application of new techniques have caused constant re-evaluation of their scenarios. Little more than a hundred years ago, the role of glaciation in shaping the land mass was not understood. The use of radiocarbon dating and the analysis of deposits of pollen and microscopic fossils to determine changes in plant and animal life (and, by implication, changes in the environments of these organisms) have become established procedures only fairly recently. Even as late as the early 1960's, the concept of continental drift, now basic to understanding the relationships of the great land masses through time, was ridiculed by many serious workers. Hence, our attempt in the next few pages to fit together a coherent pattern of events can only be read as an interim report, subject to revisions.

For the Northeast, a convenient time to begin the act of reconstruction is at or near the close of the last glaciation, when most of the area was buried under ice. For years, scientists believed that the glacial periods were relatively brief, each episode of bitter cold being followed by long intervals of warmth. Now it seems clear that this was not the case, that indeed the reverse was true: the cold persisted for intervals lasting about a hundred thousand years, separated from one another by brief respites. The last major glacial episode began in eastern Canada; a growing ice sheet resulted from a widespread lowering of the snow line. As snow cover increased, more of the sun's radiation was reflected, the atmosphere began to cool, and increasing amounts of precipitation reached the earth as sleet or snow. The ice sheet grew southward, reaching its maximum development about 20,000 years ago, when it extended as far south as Long Island Sound and westward across Pennsylvania, Ohio, and much of the Midwest before swinging north across the upper sections of Montana,

Idaho, and Washington. Near the ice a zone of permafrost, in which the soil never thawed beneath its surface throughout the entire growing season, supported tundra vegetation. Feeding on the lichens and mosses here were arctic forms of wildlife, including lemmings and musk ox, whose remains may be found today as far south as Illinois and Pennsylvania.

At the height of the last glaciation, much of the Northeast lay beneath the ice. The tree species which form our modern forests survived in various areas far to the south.

As the ice sheet formed, increasing amounts of the earth's limited water supply were bound up in it in the form of ice. As a result, at the height of the glaciation sea levels were reduced as much as 150 feet below their present mark. The ice sheet itself covered much of the former seabed in northern New England; to the south, however, a vast expanse of newly created land was exposed.

For years a major debate among ecologists centered on the degree to which vegetation migrated southward toward the equator as the ice advanced. E. Lucy Braun, writing in 1950 the classic *Deciduous Forests of Eastern North America*, concluded that "Fossil and vegetational evidence are in accord, both indicating that the deciduous forest was not displaced south of the glacial boundary, except in a band of varying width along the ice front. . ." Rebutting her views, E. S. Deevey wrote that while ". . . it is not argued that tundra conditions held sway along the Gulf of Mexico or that climatic zones and their accompanying biota marched and countermarched with military precision from Mexico to Canada," nonetheless ". . . taking all the facts together it seems . . . that glacial chilling in the southeastern States must have been fairly extensive and that the warmth-loving species . . . survived in peninsular Florida and in Mexico, and have subsequently migrated to their present localities."

Now, nearly three decades since these arguments were formulated, increasing numbers of pollen samples taken from bogs below the glacial boundary have shown that the deciduous forests were displaced very far indeed — so far that where the tree species survived the glaciations is not entirely clear. Tundra vegetation, while fairly limited, occurred at least as far south as Maryland. The coniferous forests occupied a broad band below the tundra and moved out onto newly available sites on the open seabed. Temperatures are believed to have been depressed an average of 15 to 18 degrees Centigrade over much of the United States, and even the southern Appalachian highlands (long believed to be "refugia" for the beleaguered hardwoods because of the rich assortment of deciduous trees which grow there now) supported for the most part boreal coniferous forest.

As the ice sheet began to melt — and it did so unevenly, with periods of retreat interrupted by standstills and even surges where the glacial ice moved forward again — it left behind piles of debris, jumbled rocks of various sizes, sands, silts, and clays all intermingled. As the limit of the ice moved northward, vegetation followed it. The tundra, closest to the ice, passed through the Northeast, leaving behind a set of plant species whose descendents persist in the alpine zones of the Presidential Range, on Katahdin, and to a lesser extent on other mountains of the region.

In Connecticut by 12,500 years ago, an open spruce wood-land had replaced the tundra, with lichens and mosses carpeting patchy open spots between the trees. Animal fossils from this period are uncommon in the Northeast; mammoths, mastodons, and caribou are believed, however, to have been the most important forms of animal life. At this time, the fauna of the entire North American continent underwent a remarkable series of changes. Many of the larger species, including the mammoths and masto-dons, giant ground sloths and camels, horses and sabre-toothed cats, rapidly became extinct; the cause of their demise is still uncertain. Some authorities have attributed the extinctions to the climatic changes that occurred at the end of the glaciation. Other, more recent workers have pointed out that similar extinctions did not occur at the end of previous glaciations, that plants and smaller animals did not, so far as is known, become extinct in as great numbers as the larger species, and that marine mammals show no similar declines throughout that period. Humans, these workers argue, appear in the fossil record in North America at this time, and comparable extinctions on other continents — for the decline of the large mammals was a worldwide phenomenon — occurred not at the close of the last glaciation but very near the period when the first records of man, spreading out from the African highlands where he apparently evolved, are found in the fossil record.

If humans were responsible for this wave of extinctions, which was even greater than that which occurred with the Euro-pean settlement of the continent, how did they, with their limited technology, accomplish such destruction? Almost certainly, a number of aspects of human activity were involved. The extinc-tions may not have occurred all at once over the continent, but instead could have been concentrated at the edge of mankind's advance as he crossed the Bering Strait and fanned out across the North American land mass. Excessive hunting, exacerbated by wasteful, inefficient methods may have been part of the explanation; for instance, whole herds of animals were deliberately stampeded off cliffs to provide meat for small bands of hunters. Competition with large predators for food sources and increased agriculture, which could have disrupted the feeding patterns of browsing and grazing species, may have led to a general change in the habitat which might have favored one of these herbivores, allowing it to effectively out-compete other species. Any or all of these factors may have played important roles in what has sometimes been described as "overkill."

During the period the large mammals were disappearing for whatever cause, the coniferous forests began an ever more rapid movement northward. About 11,000 years ago, our observer fixed in space atop Mount Tom would have seen spruce filling in the gaps in the open woodland in company with alder, fir, and jack pine. The exact chronology of change varies from place to place.

In northern New England, the land along the shoreline had been shoved down below the level of the sea by the weight of the ice sheet. The ice sheet itself rested on top of the land mass; as it melted the sea rushed in, covering areas west of the present coast. In southern New England and south along the Atlantic seaboard, the land was not depressed and ocean water remained farther to the east than it is now, providing the species which had survived the glaciation there with an environment from which at least some of them could begin a migration back into their former range. Large river valleys may have served as sheltered sites for the migrations of species such as willow and aspen. Lakes, filled with meltwater from the ice sheet that was held back by piles of glacial debris, formed and in many instances later disappeared. One of the largest of these, Lake Hitchcock, which stretched from central Connecticut north to what is now Lyme, New Hampshire, drained during this northern expansion of conifers at a time very near to the abrupt appearance of human hunters in the area. These Paleo-Indians might have, from the crest of Mount Tom, witnessed the extraordinary events which surrounded the rupture of the natural dam and the rapid emptying of the waters of the lake. They might have even pursued caribou, one of the large game species which survived the Pleistocene extinctions, across the eroding sediments of the valley floor, although the drifting sands in the bed of old Lake Hitchcock were soon to be anchored in place by vegetation, while the caribou were to follow the tundra north.

By 10,000 years ago, temperatures had reached their modern levels, and the climate continued to grow even warmer. Following the surge of ocean water inland, the coastline of northern New England has remained relatively stationary, the rise of sea level caused by water entering the ocean from the still-melting ice sheet balanced by the rebound of the land. At about 10,000 years ago, however, the land mass of northern New England began to rise faster than the sea level and areas which had been previously submerged were exposed for the first time since the ice sheet rode down over them. The vegetation which moved out to colonize these sites may have been a flank of the spruce-fir forests in their northward advance; at sea level, these two species find their southern limits of distribution today on the coast of Maine, where their survival depends on the influence of the cooler maritime environment. Forests of spruce and fir were replaced at the lower elevations and latitudes by other tree species which invaded at different times, from different directions, and at different rates.

South of the vicinity of Boston, the rising sea, not counterbalanced by rebound, began to drown out the vegetation on the exposed continental slopes. White pine, which apparently survived the glaciation on the exposed mid-Atlantic seabed, migrated north and westward. Jack pines, which were associated with the spruce-fir forests, moved with these conifers up into Canada, leaving behind small, outlying stands in northern New England. Oaks increased with white pines to become, by 9,500 years ago, the most prominent species in the forest of southern New England and New York, where small, mobile bands of Indians were now established residents.

Up to this point we have discussed the forests largely in terms of their more important trees. It might be well to stress here that they are actually communities of plants, animals, and fungi (which some biologists would place in a separate kingdom of their own). These organisms are linked by an intricate series of relationships with one another and with the nonliving components of their environment. We discuss the concept of the community in more detail in Chapter 4. The idea is important in the context of the changes following the glaciation since, as the individual tree species of the community changed in abundance, the other organisms must also have been affected. Hence, as the tundra and coniferous forest were replaced in time by pine-oak woods at any given site, so also were the surviving large herbivores such as caribou replaced by moose and elk and then by white-tailed deer.

A coniferous forest in winter. / G. Bellerose

Only recently have studies of the pollen record indicated that Deevey's intuitions were correct, that "climatic zones and their accompanying biota" had not, despite the drastic changes which accompanied the glaciation, "marched and countermarched with military precision from Mexico to Canada." Instead, the communities seem to have assembled themselves, species by species, supplanting the species which moved northward. Hemlock, for example, traveled more slowly than the oaks and arrived 500 to 1,000 years later at most sites; its shade-tolerant seedlings invaded the forest floor beneath pines, oaks, birches, and maples, which the mature hemlock trees would later join in the canopy. Beech, moving from another glacial refuge, migrated parallel to the Appalachian Mountain chain at a rate of up to three hundred meters per year to take its place in the hemlock-hardwood forests.

While the major expanse of coniferous forest took a position as "taiga" below the tundra in Canada, a remnant remained behind at higher elevations. The warmer the climate at sea level on the eastern seaboard, the higher up the zones of spruce must have begun. Now, spruce and fir dominate woodlands above 5,000 feet in elevation in the Great Smoky Mountains, above 3,500 feet in the Catskills, above elevations ranging from 2,400 to 3,000 feet in the White Mountains, Green Mountains, and Adirondacks, and only 500 feet on Katahdin in northern Maine. The separation between coniferous forests above and deciduous forests below is related to the shortened frostfree period, shallower soils, and increased occurrence of clouds, fog, and hoar frost at the upper elevations. In the Green Mountains and elsewhere, a distinct zone of transition exists where neither spruce and fir nor deciduous species such as sugar maple, beech, and yellow birch are well established, where instead brushy thickets alternate with fern-rich openings. As the climate has warmed or cooled at intervals through the last 10,000 years, the critical conditions themselves and the altitudinal limits they impose must also have varied.

The climate, for example, warmed to what is known as the "thermal maximum," a period lasting from 8,000 to 5,000 years ago, with higher temperatures and in some areas less precipitation than now. At that time white pine and other species moved north of their present range. Toward the end of this thermal maximum slow-moving hickory arrived in Connecticut from its glacial refuge, which was not on the exposed seabed on the eastern coast but apparently somewhere to the south and west of our region. In the Connecticut Valley, game species such as deer, wild turkeys, squirrels, and migratory birds, as well as fish, were as abundant as they ever had been, and human populations, less mobile than before, reached their greatest densities prior to the European settlement. The Indians established homesites on terraces of the Connecticut and along the River's major tributaries, avoiding the meadowlands on the old bed of the glacial lake.

About 4,000 years ago, as the thermal maximum ended and the climate became cooler, the numbers of Indians declined and then, as their culture increasingly turned to agriculture, began to swell again.

In the Connecticut Valley at the time of the first sustained contact between Europeans and native Americans, an elaborate annual cycle of Indian activity had evolved. In March, families grouped at the major falls on the Connecticut to fish the spawning runs of alewife, shad, and salmon. During April and May, cornfields were prepared and seeded, chiefly on the floodplain where soils were light and easily hoed. Plantings were probably limited in scope because the kind of agriculture practiced (a method sometimes known as "slash and burn," in which fields are cultivated until the soil is exhausted and then abandoned until, in the course of time, native vegetation redevelops and fertility is restored) requires intensive labor and a large amount of land, only some of which is productive at any given time. Fishing and hunting of waterfowl and mammals continued through the summer, but families tended to be concentrated near the fields where the crops of corn, pumpkins, tobacco, beans, and squash ripened by August. As autumn approached, wild plants were gathered and stored: herbs, nuts and berries for foodstuffs, reeds and rushes for basket making. After the harvest, small groups of hunters moved away from the main settlement into lodges on the major tributaries for a period of hunting. Then, from December through March, the tribes reassembled in their major settlements, subsisting on stored foods, hunting, and fishing through the ice of the frozen streams.

On the coast of northern New England rebound no longer exceeded the rise in sea level. Indeed, over the last 3,000 years, the sea has risen by a few feet along the entire coast; only about 300 years ago did the level of the shoreline become stabilized.

At just about the time the sea stopped rising, a small event foretold a larger pattern of displacements. In 1602, during Bartholomew Gosnold's explorations south of Cape Cod, twenty-six cedars were cut on the morainal island of Penikese and shipped to England for sale; the logs were intended, by contemporary accounts, to "make masts for the greatest shippes of the world: Excellent timbers of Cedar and boords from curious building. . . .' Gosnold's men also stole an Indian canoe hidden above the beach. Writing in 1874 after a summer spent on Penikese, D. S. Jordan described the then treeless island as "about as barren looking a pile of rock and stone as one could well imagine. . . ." Since the end of the last glaciation, the changing climate had directed the course of change in the northeastern landscape; since the time of Gosnold's landing, another dominant force has come into operation — man.

3.

The First Europeans

American literature only hints at the effects of European man upon what was seen as a largely unspoiled country. For example the Pilgrims, landing near what is now Provincetown on Cape Cod, found the area "all wooded with oaks, pines, sassafras, juniper, birch, holly, vines, some ash, walnut: the wood for the most part open and without underwood, fit either to go or ride in." Henry David Thoreau, visiting there in the mid-nineteenth century, found "scarcely anything high enough to be called a tree" and accused the Pilgrims of exaggeration. Unravelling the truth from accounts of this period is difficult; concerned with the difficulties of daily life, too few early settlers left detailed accounts of the nature of the countryside they were adopting for their homeland. As the colonial towns and villages increased in number, the Indians retreated, often abandoning their own habitations well in advance of the white man's spread. In the absence of cultivation, many fields grew up to second growth, often before the colonists arrived. Moreover, where burning had been a factor in determining the open, park-like nature of the woodlands, dense brush flourished in the interval before resettlement. Hence, what the colonists saw was in many instances not the original landscape, as it had been at the time of the first explorations, but rather a landscape which reflected the recent flight of the Indian populations.

By the mid-eighteenth century, open conflicts between the Indians and the colonists had largely ceased, and over large areas the forests were cleared for cultivation or pasturing. Above the glacial boundary, boulders left scattered by the retreating ice were heaped to form stone walls which divided the land into a mosaic of parcels that, depending on their owners' inclinations, might be plowed or grazed, burned or allowed to grow up again in second growth. The white pines, particularly the largest specimens reserved for use as ship masts, were the first heavily exploited trees. Human activities soon provided a new role for the pines; rather than maintaining themselves as scattered giants towering over a lower canopy of hardwoods, the pines became important invaders of cleared and abandoned pastures. Avoided by grazing cattle, they formed dense stands in the increased light now available at ground level. White pine stands, if clearcut, quickly revert to stands of deciduous trees, a fact which Thoreau was one of the first to note. (See Chapter 5, "Natural Succession," and Chapter 6, "Man and the Process of Succession," for a discussion of some of the factors involved here.)

The pitch pine forests which grew on sandy soils, particularly in the southern portions of the region, were cut for tar, turpentine, and lampblack. Later pitch pine would be an important fuel for locomotives.

Oak and hickory were also used for fuel where the land was not completely cleared. After cutting, several shoots developed from each stump, maintaining the same sprout hardwood type of woodland which had developed earlier because of fire.

In southern New England, three-fourths of the land was cleared by the early half of the nineteenth century. On the Berkshire plateau, settlements arrived later. About half of the land was pasture, but a substantial portion was never occupied. The coniferous forests, which had supported few Indian inhabitants, were the last to be settled. Agriculture was light here and logging the principal disruption. In the Catskills, hemlock was cut at the lower elevations for tanbark; heavy logging began there and in the other mountainous areas in the 1800's.

The destruction of the forests of the region was followed after the Civil War by a gradual cessation of cultivation and pasturing and a reversion of this land to forest. We describe this process in Chapter 6 and discuss in Chapter 25 the relation of agriculture to the world food problem. Of importance here is the fact that the returning forests tended to contain the same tree species as before, with those species now at different levels of abundance. Walnut and hickory showed a general decline. Because of its ability to rapidly rejuvenate from sprouts, American chestnut increased and then, within this century, has disappeared as a dominant forest tree because of chestnut blight, a disease appearing as recently as 1910.

In the Catskills at lower elevations, beech and hemlock decreased while red oak, chestnut oak, and sugar maple became more prevalent. In northern Vermont, because of its slow rate of regeneration, beech, which according to surveyors' records had covered forty percent of the land before settlement, declined to five percent or less, while second growth species such as pine, hardhack, and poplar increased.

Animal populations, so severely disrupted at the end of the last glaciation, showed a series of changes following the European settlement which in some ways parallels those which occurred when man first appeared on the continent. As we have seen, one of the motives ascribed to the Indians of the pre-colonial period for burning forests was to improve habitat for such game as white-tailed deer, wild turkeys, and heath hens. If we examine these three alone, we can contrast the differing fates which a species can meet as it adjusts or fails to adjust to a set of new conditions.

The heath hen, affected most drastically, is now extinct; the birds once ranged from southeastern Maine down the coast as far as Virginia in open scrub oak and pine habitat which was in part maintained by fire. After burning of the woods ceased, their numbers declined as the land reverted to other types of vegetation. Overhunting by man, possible competition from ruffed grouse and quail, and predation by foxes, raccoons, and feral cats are among the factors which interacted to exterminate the heath hen from the mainland by 1869, despite a number of weak legislative attempts to save it. (In 1831, for instance, when the bird was already extremely rare, the Massachusetts legislature passed an act to protect it during the breeding season only. Six years later, a closed season on heath hens to last four years was declared. The closed season was later extended with the provision that by vote of town meetings individual communities might exempt themselves from it.)

On Martha's Vineyard, the home of the last surviving population, the town of Tisbury alternately allowed or forbade hunting for heath hens for half a century "under periodical juggling of the statutes," according to Edward Howe Forbush, whose third volume of *Birds of Massachusetts and Other New England States,* published posthumously in 1929, contains the statement "There may be two or three specimens of the Heath Hen still surviving on Martha's Vineyard while this is being written, but probably there will not be a single individual living in the entire world by the time this volume reaches the public." The last sighting of a heath hen actually occurred in 1933, despite the fact that a reservation was set aside early in the century for it on Martha's Vineyard. Although their numbers there had increased with protection to perhaps as many as 2,000 birds, a fire on the breeding grounds and subsequent predation by cats and goshawks reduced the population to a point from which it could never recover.

A white-tailed deer near Gorham, New Hampshire 1905, / AMC photo collection

The white-tailed deer also declined following the European settlement. As we note in Chapter 6, Thoreau found the report of a single deer near Northboro worth recording in his journal. Deer are best adapted to second growth and forest edge. They thrived in the burned-over woodlands of the pre-colonial period and, with protection, they are thriving again as forests re-establish themselves in abandoned fields and pastures. We will in Chapter 26 make a distinction between renewable and non-renewable resources. The heath hen is an example of a potentially renewable wildlife resource now lost forever through extinction; the white-tailed deer, subject of numerous studies and attempts at management, has been maintained.

Seen on a national level, the contrast in these two species demonstrates even more strikingly the degree to which species differ in the way they are affected by man's activities. The heath hen is considered by many scientists to have actually been a race of the more western prairie chicken. Prairie chickens have, in fact, on occasion been released in the East and later mistaken for their extinct relatives. The prairie chicken survives in two races, the endangered Atwater's prairie chicken, which is found only in a limited area on the Texas coast, and the greater prairie chicken, with a wider but still contracting range. As a group, prairie chickens have done poorly since the coming of Western civilization. White-tailed deer, on the other hand, with a population estimated at 12,500,000 animals in 1978, are more abundant now than they were in 1600.

The wild turkey, the third of the species the Indians may have managed as a resource by deliberately firing the woodlands, was neither as successful as the white-tailed deer nor such a failure as the heath hen. Reported at its northern limit in southwestern Maine, it increased south over much of the eastern half of the continent in deciduous woodland habitats. Wild turkeys were formerly abundant in Massachusetts, Rhode Island, and Connecticut. They extended up the valley of the Connecticut River into Vermont and north in New Hampshire to Lake Winnepesaukee.

Early colonists and explorers commented on the changing status of the birds, Captain John Smith reporting in 1622 "great flocks of turkies" and Thomas Morton, ten years later, corroborating " . . . turkies there are, which divers times in great flocks have allied by our doores . . ."

William Woods, in the same period, observed the birds could " . . . runne as fast as a Dogge, and flye as well as a Goose." Within fifty years, however, Josselyn commented " . . . tis very rare to meet with a wild *Turkie* in the Woods." He attributed this to "the *English* and the *Indians* having now destroyed the breed." By the end of the eighteenth century most wild turkeys in Massachusetts occurred along the Connecticut River and farther west. The last turkey was seen in Connecticut in 1813; the birds were still surviving in Vermont in 1842.

Amherst professor Edward Hitchcock, for whom the glacial lake was named, reported wild turkeys as being frequently seen on Mount Holyoke in 1835; the birds persisted there and on adjacent Mount Tom through the middle of the nineteenth century. Indeed, had our survivor, whom we imagined on Mount Tom since the last glaciation, managed to remain in place, he might have seen one of the more pathetic minor episodes in the history of wildlife since the European settlement. Those who hike the Metacomet-Monadnock Trail today know that Mount Holyoke and Mount Tom form a long, east-west running ridge, the Connecticut River separating the two ranges. Shortly before the outbreak of the Civil War, a group of hunters from Springfield and Holyoke, Massachusetts, divided themselves into two parties to perpetrate a remarkable demonstration of the cunning and lack of foresight of which humans can be so capable. The first group placed itself in ambush on the east face of Mount Tom; the second crossed the Connecticut River and ascended Mount Holyoke to drive the remnant flock of turkeys westward toward the stream. The birds, having retreated to the end of the range, took flight across the Connecticut and were met and destroyed by a volley of fire from the hidden sportsmen on Mount Tom. Shortly afterwards, as a result of similar persecutions everywhere, wild turkeys were gone from all the eastern seaboard.

Their saga, however, does not end as sadly as that of the heath hen. The species survived in other areas and has been reintroduced, although with some difficulty, into its former habitat. In some locations it is now re-established and spreading, saved by the use rather than the misdirection of human intellect.

We have discussed these three examples because the species were probably beneficiaries of man's activities in the precolonial woodlands. If one looks systematically at the fates of animal populations (a good source for mammals is the recent *Wild Mammals of New England* by A. J. Grodin) one can discern a number of patterns of success or failure since the European settlement.

Extinction in the final sense — i.e., elimination of a group throughout its entire range — has been more frequent with birds than with mammals. In addition to the heath hen, the passenger pigeon, great auk, and Labrador duck, all of which formerly inhabited the region, now exist nowhere on earth. The sea mink is the only mammal within our area which has, to our knowledge, met this fate within historic times.

Numerous species have, however, disappeared from the

area while persisting elsewhere on the continent. Among these are the wolverine, the American elk or wapiti, the eastern woodland caribou and the once-common eastern timber wolf, which Thomas Morton, himself an unwelcome neighbor of Plimoth Plantation, describes as ". . . fearfull Currs . . ." which "pray upon the Deare very much." Wolves were known even in settled areas until the time of the American Revolution. They declined over much of New England by the end of the eighteenth century, persisting for a time in the mountainous, less populated regions.

Another great predator, the eastern mountain lion, was also found throughout the region; not as common as the wolf, it was also eliminated. Like the wolf, the mountain lion was considered a dangerous pest and was hunted for bounty; unlike the wolf, it is believed by many to be making a return — without man's assistance — as the forests are restored and populations of deer, its natural prey, increase. The elk and the caribou have been deliberately reintroduced, but not with much success.

Other species, including the lynx, the marten, and the mink, persist in the area but were more broadly distributed previously, having retreated to the western and more northern regions. As in the case of the white-tailed deer, not all reductions in animal populations have been permanent. The porcupine, black bear, fisher, and beaver all declined and are now, at least in some areas, more common than before the settlement; this is because of increased second-growth habitat, controls on hunting, and in some cases repeated deliberate reintroductions. These and other groups may be taking over niches once occupied by the species which have declined.

It is surprising to realize, when we take stock of animal populations now living in the area, that there are a number of creatures living here which were absent in 1600 or before. (A similar situation exists in the plant kingdom but is beyond the scope of this discussion.) Some of these are native to the North American continent and are expanding their ranges north or east; the Virginia opossum, gray fox, cottontail rabbit, and coyote are examples, as are the cardinal, mockingbird, tufted titmouse, and turkey vulture. Others have been introduced by man — either deliberately or by accident, as in the unfortunate examples of the house mouse, Norway rat, starling, and English sparrow. Some deliberate introductions, including the European red fox and the ring-necked pheasant, have contributed seemingly permanent new residents to some parts of the region. Others, including the European hare, fallow deer, blacktailed jackrabbit, and European wild boar, were brought in as potential game species, but remain limited in their distributions.

There is a fascination in pursuing this theme of change, and we return to it in several later chapters. Forests decline and then recover to be in some ways similar and in some ways profoundly different from their former states. Animal populations prosper or decline, invade new territories or disappear entirely. For the first time in our history, man as a species has begun to assemble sufficient knowledge about the world we occupy to develop insights about not only the past or present but even what might be expected in the future. There is no question that much of what we have done is unfortunate; a number of later chapters will address these problems in detail. It seems best, however, to begin with the simple hope that as *Homo sapiens*, thinking man, we may soon attain sufficient grace that our actions will be less disruptive in the future, so that we may avoid contributing to the destruction, not only of ourselves, but of our region and indeed our planet.

4.

The Ecological Perspective

Ecology has come to mean a variety of things to different people. To some it denotes a way of life in mystic communion and balance with the laws of nature; to others it simply represents another in a series of fads and preoccupations which range from the foolish to the sinister.

Indeed, this confusion regarding the meaning of the word has led some scientists to urge their colleagues practicing ecology to find some other word for what they do. Nonetheless, it is impossible to discuss the relationships of man with his environment in the Northeast or elsewhere without some reference to ecology, so for the purposes of this book we will use the word in two rather specific ways.

The first and classic use of the term considers ecology to be the science which investigates the relationships between organisms and their environment. The second, more careless, usage refers to the relationships themselves. Thus an ecologist may study the relationships of hemlocks or white pines or evening grosbeaks or whatever living things he chooses to factors such as light, temperature, moisture, wind, soil types, or other plants and animals. He might well describe these relationships as "the ecology" of hemlocks, pines, or grosbeaks and, if he is looking at populations of these creatures, as "the ecology" of pine or hemlock forests or of grosbeak flocks.

One of the most important ecological concepts to arise in recent years is a realization that organisms — or groups of organisms — cannot be considered outside the matrix of their total environment, the complex of living and non-living components said to form the *ecosystem*. Hence, the ecologist who began by studying only white pines will have, at some phase of his work, to think of each pine as a member of a white pine population. This population, he will realize, occurs with populations of other kinds of plants and animals — perhaps invading hemlock seedlings on the forest floor and migratory flocks of seed-eating evening grosbeaks, along with a myriad of other species: herbs, shrubs, fungi, soil-dwelling bacteria and protozoans, ruffed grouse, squirrels, warblers, or whatever — as members of a *community*, a group of populations with mutual relationships to one another and the nonliving parts of their environment. As simple as the ecosystem concept sounds, too many of our current environmental problems have arisen because we fail to realize how complex each ecosystem really is.

On the left, caterpillars feed on leaves in the first step of a food chain. On the right, a blackheaded grosbeak eats a monarch butterfly, which the caterpillar has become, while in the background a Cooper's hawk, a tertiary consumer in this food chain, attacks another grosbeak.

How big is an ecosystem? The earth itself can be thought of as one. Quite often, however, it is more useful to apply the concept to much smaller units: the forest ecosystems of the Northeast, the salt marsh ecosystems of the coast, the alpine ecosystems above timberline. These broad ecosystem types are sometimes known as *habitats* — the hemlock-hardwood forest habitat, for example. Indeed, in our own research we have found it useful to consider as ecosystems even smaller units within habitats: a single pond or river, perhaps, or a single spruce-fir zone on a given mountain. Within any ecosystem, regardless of its size, certain basic sorts of relationships may be observed. One of the most important of these is the presence of *food chains* — or more properly *food webs* — interacting groups of organisms, some of which normally make food, fixing the energy of sunlight, others of which eat all or part of the energy fixers, and others of which ultimately break down the food makers and the food consumers to non-living, inorganic forms.

On land, the most important food-making organisms, the *producers*, are the higher plants, plants with true stems, roots, and leaves — our white pines and hemlocks are good examples. In the oceans, lower plants — algae of various sorts, both microscopic plankton and the heavy-bodied kelps and rockweeds of the rocky coast — are most important.

Consumers fall into several groups: herbivores or *primary consumers*, the evening grosbeak shredding hemlock cones or the deer browsing white pine growth; *secondary consumers*, the owl which seizes the unwary grosbeak or the hunter who has replaced the mountain lion as a predator on deer herds in the region; *tertiary consumers*, the hawk which eats the warbler which has fed on insects which have browsed on white pine buds. Even higher levels of consumerism exist, and a consumer may not confine his diet to one level of prey — our human hunter may eat his venison with parsnips and a lettuce salad.

The fungi and bacteria which break down the bodies of producers and consumers alike are known as *reducer organisms;* these are important members of the ecosystem which are very poorly understood.

Simple and obvious as the idea of food webs may seem, several of our most pressing recent environmental difficulties have resulted from failure to understand the intricacies of this simple energy-exchanging relationship within the ecosystem. A bitter example are the pesticides which, applied to plant life to destroy grazing insects (primary consumers), have killed not only target organisms but also secondary consumers which might have kept the herbivores in check in the absence of man's intervention. Some of these pesticides, fat-soluble, accumulate and become more concentrated in the consumer portions of the food web. Hence, the hawk which captures the warbler feeding on insects previously exposed to DDT (applied by man, perhaps to maintain a trim and sightly white-pine grove) may, at nesting time, lay eggs which fail to hatch because of abnormal weakening of the shell. Ospreys have recently showed a marked decline along our coast, and the peregrine falcon no longer breeds on the ledges of Mount Tom in western Massachusetts. These birds are victims of man's interference with the food web, and they are not the only instances.

Food webs in tundra or alpine regions are notoriously simple and vulnerable to disruption. One of the most sinister effects of man's interference has been observed in both the New and Old World arctics in the past two decades. The most important producers in these regions are lichens known as reindeer moss. The lichens are browsed by caribou in the North American tundra and by reindeer, the caribou's close relation, in the Old World. Other producer organisms — grasses, sedges, and dwarf willows — support such primary consumers as lemmings, snowshoe hares, and ptarmigans. Secondary consumers include the arctic fox, snowy owl, wolf, and man himself. In the summer, the web becomes more intricate as migratory birds return and insects emerge from dormancy. In autumn the web contracts. Man's role in an ecosystem as simple as this one is precarious indeed; the subjects of the famous documentary *Nanook of the North,* for example, perished from starvation during the period when the film depicting their interesting and difficult lives was being shown in theaters across America.

The above-ground nuclear testing of the 1950's added a new kind of threat to residents of arctic ecosystems. Unlike higher plants, which absorb nutrients through their roots from the soil, reindeer mosses acquire minerals directly from the dust that falls on their intricately branched plant bodies. Nuclear fallout contains several radioactive components which the reindeer moss

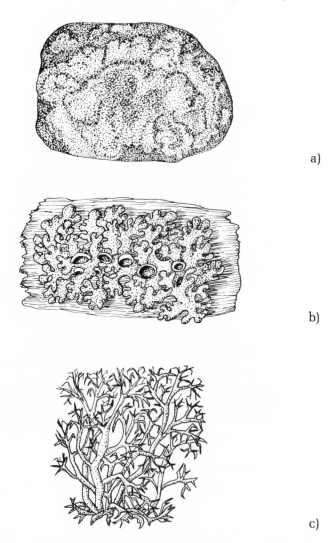

a)

b)

c)

"Reindeer Moss" is a common name for a species of lichen which is the most important producer in the winter arctic food web. Lichens are composite organisms which result from the interaction of algae and fungi in a special relationship known as mutualism. Both partners in mutualism benefit. The algal cells carry on photosynthesis and provide food for the fungi. In turn the fungi absorb water and nutrients and provide the algae with support. Lichens may be either (a) crustose and tightly pressed against rock or tree trunks or (b) leafy, less firmly attached forms or (c) shrubby tufted forms such as reindeer moss. Crustose and leafy lichens are important in the process of primary succession on rock.

absorbed; these were then picked up by browsing caribou and reindeer. One of these radioactive components, strontium 90, is chemically similar to calcium, tends to concentrate in milk, and is ultimately stored in bone, where it may affect the cells of bone and marrow. Another component, cesium 137, accumulates in muscle tissue. Eskimos, Laplanders, and other residents of arctic areas for whom caribou or reindeer meat is a major food source have been shown to have far greater concentrations of these elements in their bodies than do people living in the temperate zones.

Lichen. / D. A. Haskell

      The ultimate damage done as a result of nuclear testing may be a genetic one not evident for generations. What is important for us now to realize is that the addition of substances to the ecosystem is a serious, potentially disastrous activity with far-reaching effects we may well not be able to predict at present. It should be done, if at all, with the utmost caution and respect for what may be destroyed. We ourselves, at the apex of the consumer line of many food webs, may be in the end our own victims.

5.

Natural Succession

Henry David Thoreau, speaking to the Middlesex Agricultural Society in Concord during September of 1860, described how, in his work as a surveyor, he was often asked "how it happened that when a pine wood was cut down an oak one commonly sprang up and *vice versa.*" Thoreau detailed methods by which seeds might be transported, pines and maples by wind and water, acorns and other nuts by animals. The shade of a dense pine woods was less favorable for the growth of young pines, he claimed, than for young oaks, the seeds of which were carried in by squirrels. His essay, published under the title, "The Succession of Forest Trees," was one of the first attempts to deal with a fact observable to all who spend any period of time outdoors: the composition of communities in most ecosystems changes, often so drastically that one community type seems to replace another in a sort of relay which continues until a final, relatively stable stage, often known as a *climax community,* exists.

Few aspects of ecology are so controversial as the mechanisms involved in what Thoreau referred to as *succession.* Some ecologists, indeed, would abandon the term succession and the idea of the climax community entirely for phrases such as "vegetational development" and "steady state." Others view succession as one of the great and fundamental concepts of their science. Regardless of the debate, change in ecosystems does occur and on such a regular basis that those dealing practically with environmental problems must have some understanding of its course.

For convenience we may distinguish two basic types of successional development: *primary succession*, which begins in the absence of soil and which involves either the breakdown of rock or the filling in of a body of water to provide a surface wherein higher plants may become established, and *secondary succession*, which goes on in areas where soil is present but from which all higher plants have been removed.

Most of us have observed at one time or another patterns of vegetational development on boulders and rocky cliffs. Thin layers of gray-white crustose lichens invade bare rock; larger fleshy-bodied lichens overlie the crustose types where a layer of soil has formed from broken-down rock material, decaying lichen bodies, air, and moisture. Mosses, which lack true stems, roots, and leaves, replace the lichens and are themselves replaced by higher plants — ferns, grasses, sedges, and shrubs such as low-bush blueberry, whose tough roots penetrate rock crevices and contribute to a deepening of the soil and in time to their own replacement. At present this sort of sequence may seem relatively unimportant in northeastern woodlands; nonetheless, this process and the soil-building which result from it have made our forests possible.

The other basic form of primary succession, the filling in of bodies of water through a process which involves *eutrophication*, has been accelerated recently by man; we will discuss this in a later chapter along with other problems connected with water resources.

Secondary succession, the development and change of communities in areas where soil is present but from which higher plant life has been removed, may begin after fires or major storms — hurricanes or strong northeasters in our region — or after man has cleared the land for fields or pastures. We will consider for the present only successional developments which occur without man's intervention; in order to fully appreciate these, we must first attempt to understand the climax or steady state ecosystems in our area. Unfortunately, this is not an easy task; stands of virgin forest are few and far between and those which occur often have persisted in relatively inaccessible — and hence atypical — areas. All will agree that the climax or steady state was almost certainly a forest (just as in the arctic it is tundra and in much of the American West a desert); the question is: what kind of forest? New

England and upstate New York, as we have seen, are areas characterized by a transition between coniferous evergreen forests to the north and deciduous hardwoods to the south. The composition of

This old sugar maple is one of a line of trees which were incorporated in a stone wall at the time the adjacent land was cleared for cultivation. Farms in the area are now abandoned. A survivor of the original forest, the tree is now surrounded by vigorous young sugar maples which became established when the fields were no longer cultivated. Secondary succession is occurring here and the land is returning to its original forested condition. / C. J. Burk

the forest one finds in any given section depends upon elevation, the slope or lay of the land, historical factors, and various other conditions, as well as the degree of latitude and climate. Hence, in the Green Mountains of Vermont a recent study has shown that a northern forest type dominated by balsam fir, red spruce, and white birch occurs above 2,600 feet, while at lower elevations sugar maple, beech, and yellow birch predominate. The line between these forest types is not sharply demarcated, consisting of a band of forest where species from both major habitats occur mixed with one another. On the east-west running Holyoke Range in western Massachusetts, naturalists have long noted differing forest types on the south- and north-facing slopes. Thus, if we hike the Metacomet-Monadnock Trail from the Notch to Mount Nor-wottock and scramble down past the Horse Caves, we have a choice of two routes back to the starting point: a bushwhack on the north side through hemlock, beech, and sugar maple or a ramble on old wood roads on the warmer, drier south-facing slopes through what is largely oak woods.

Forest communities are complex and diverse far beyond the species composition of the trees which dominate them, so complex, indeed, that few workers have attempted to account for all the kinds of organisms present. One of our favorites of these rare attempts was made by Victor Shelford in his book *The Ecology of North America.* Shelford estimated that in a hypothetical ten square miles of deciduous forest, the approximate area normally occupied by a city with 80,000 to 100,000 inhabitants, one would expect to find the following:

— 750,000 trees 3 inches in diameter at breast height or larger — oaks, maples, basswood, hickories, walnut, formerly chestnut

— 786,000 tree seedlings

— 2,810,000 shrubs — spicebush, paw-paw, etc.

— 230,000,000 to 460,000,000 herbs — bloodroot, violets, jewelweed, nettles, wild ginger, etc.

— 26,880,000,000 invertebrates, mainly spiders and insects

— 8,960,000,000 arthropods large enough to be counted with the naked eye — snails, millipedes, centipedes, earthworms, etc.

— 7,680 pairs of small nesting birds

— 20-50 predatory birds — great horned owl, redtailed and redshouldered hawk, etc.

— 160,000 to 320,000 mice

— 20,000 to 40,000 squirrels

— 200 wild turkeys (formerly)

— 100 to 840 white-tailed deer (optimum 400)

— 2 or 3 pumas (formerly)

— 30 foxes

— 1 to 3 wolves (formerly)

— 2 bears (formerly up to 5 per 10 square miles)

— assorted other animals including voles, marmots, fish, and snakes which were not counted

Shelford has omitted from his compilation most of the reducers, the bacteria and protozoans which occur at the rate of an estimated two million or more individuals in every gram of soil. We must also ask another question: what happens when communities of this complexity are disturbed? What happens in these forests after a hurricane or a fire set by lightning? What seems increasingly clear is that within the total structure of the forest are elements — species of plants and animals — which increase and act as stabilizing members of the ecosystem during periods of stress. One of these stabilizing elements, for example, is the common pin cherry. P. L. Marks at Cornell University has shown that the seeds of pin cherry may remain capable of germination when buried in the forest soil for as long as fifty years. Tremendous numbers of the seeds (which are distributed by robins, grosbeaks, waxwings, and other birds which eat the cherry fruits and pass the stony pits through their digestive tracts) lie buried in the forest soil, an estimated 200,000 per acre in one New Hampshire forest. When these forests are disturbed, when the dominant tree species are destroyed, these buried seeds quickly germinate and young pin cherry trees form dense stands which hold the soil in place, preventing erosion and loss of the product of centuries of slow primary successional development, setting the stage for the reinvasion of the climax forest types.

A lesson to be gained from studies such as Marks's is the importance of diversity, the continued presence in an ecosystem of a variety of types of plants and animals. Few lumbermen depend on pin cherries directly for their livelihoods, and yet, within the forests on which they do depend, the cherry is essential. How many more organisms play equally important, not fully understood, roles? It would be hard to overestimate their numbers.

6.

Man and the Process of Succession

Surprising as it may seem, despite urban sprawl, suburbanization, and uncontrolled development in certain areas, much of the northeastern landscape is far wilder than it was a hundred years ago. In 1833, Edward Hitchcock, Professor of Chemistry and Natural History at Amherst College, questioned, in a catalogue of Massachusetts animals, whether the cougar, beaver, and pine martin still existed in the state. He doubted it. Thoreau, in his essay "Natural History of Massachusetts," observed that "the bear, wolf, lynx, wildcat, deer, beaver, and martin have disappeared" from the vicinity of Concord. In his journals he made special note of the appearance of a deer at Northboro and of a wildcat heard in his own neighborhood.

However, recent issues of *Massachusetts Wildlife*, published by the Division of Fisheries and Game, contain hunting regulations for antlerless deer in Middlesex (Thoreau's) County and descriptions of a "new wolf", possibly a coyote or cross between a coyote and a wolf, along with reports of increased black bear populations and the probable return of the cougar near Quabbin Reservoir and elsewhere. The beaver, moreover, has been long since re-established, and has become something of a pest in many regions.

In large part these changes are the result of a series of human events: widespread deforestation after settlement, movement of human populations from small farms into rapidly expanding cities, industrialization, the shift in dependence from falling water to steam as a power source, abandonment of pastures and croplands, and finally the regrowth, through the process of secondary succession, of much of the northeastern forest. These changes did not occur at the same time in any given area. A walk through the Conway State Forest — up the Henhawk Trail to Sinkpot Road, along Avery Brook to Cricket Hill, down Guinea Gulf and back in a loop with a stretch on Main Poland Road — resembles a stroll through a house of many rooms. These woods, fields, and farms given up at different times throughout the years contain chambers of straight, self-pruned white pines without an understory, galleries of white birch, halls of beeches, yellow-brown in autumn, a den of spruce impenetrably thick and dark, a terrace above a beaver marsh of maple swamp with scarlet foliage by late summer. The trail itself, a corridor walled by old stone fences, is lined in spots with ancient sugar maples, a museum of trees some of which are six feet in diameter. Deer are not uncommon there and bears have been reported; with luck one can walk for nearly twenty miles and not see a single human being.

By what stages did these changes come about? When a field, an old cornfield or potato patch, is no longer cultivated it is invaded quickly, certainly by the next growing season, by large numbers of *pioneer* organisms, many of which are annual plants and hence able to complete their entire life cycle, from seed to seed, in a single year. Crabgrass and ragweed are common pioneers, although almost any plant may also get established when its seeds are carried to the area. Animal life — at least those large forms we can see — tends to be scarce the first year and food webs simple: insects browsing on the pioneer vegetation, spiders preying on the browsers, field mice and cottontail rabbits and occasional marauding mourning doves and meadowlarks.

During the second and third years, annual plants decrease in abundance, replaced for the most part by longer-lived perennials which compete with them for space and nutrients. Goldenrods, asters, wild strawberries, hawkweeds, and a variety of sedges and grasses are the most important plants throughout this period, but young shrubs and seedling trees, precursors of the next stage, are almost always also there. Food webs become more elaborate and the number of animal species increases, with ground-nesting birds and extensive tunneling by field mice through the broomsedge clumps.

The garden plot behind the Brussels sprouts in the foreground was cultivated in April and then abandoned. In September, at the time this photograph was taken, the plot supported a solid stand of crab-grass, a pioneer in old field succession throughout much of eastern North America. / C. J. Burk

The same garden plot has not been cultivated for two years. It supports perennial grasses and a variety of weeds. / C. J. Burk

This pavement on the floodplain of the Connecticut River is a remnant of a drag racing strip that has not been used since the early 1960's. The area to the left has not been mowed in recent years and supports perennial herbs, including goldenrods and grasses which are typical of prairie regions to the west. / C. J. Burk

Staghorn sumac invades the grassy margin of the area seen in the photograph above. The sumacs may become small trees; they in turn will be replaced by silver maple, box elder, and other trees of the floodplain forest. / C. J. Burk

The next phase of development involves a replacement of the perennial herbs by woody species — blackberries, steeplebush, sumac, sweetfern, alders, birch, and aspen — which may form dense tangles of vegetation. A study conducted in the Catskills found yellowthroats, song sparrows, chestnutsided warblers, towhees, robins, field sparrows, and catbirds to be the most important nesting birds of these communities. The increased food and cover provided by the brush allows a further broadening of food webs. Deer thrive, and indeed the prevalence of so many fields at this state of development now accounts in large part for the increase in deer populations — perhaps to or above the level of their abundance at the time of settlement.

The final set of changes which occurs as the forest returns is the growth of the sapling trees, which appear in the brush stage, into a canopy which shades out many shrubs and perennial herbs and permits an invasion in the understory of plants which demand less light — hobblebush and mountain laurel, trilliums, bloodroot, and many of our ferns. The transition from brush to forest may proceed by various routes and may be delayed for centuries. In some areas, given an abundant seed source and enough bare mineral soil for the seedlings to become established, a stand of white pine may replace the brush. Most green plants, including young white pines themselves, cannot survive in the reduced light and heavy litter of needles beneath a canopy of pine; among those which will persist are seedlings of hemlocks and certain hardwoods — red oak, sugar maple, beech, and others — which in stunted form may endure unfavorable conditions for decades until a break in the canopy, the result of windthrow or the death of the pines from old age or logging, allows them sufficient light and space for a surge of growth to take their place as dominant species of the changing community.

If the abandoned area was a pasture rather than an old field, junipers may be the shrub invaders, forming almost impenetrable thickets of a sort most upland farmers know and hate. These thickets may in turn be invaded by seedlings of white birch which, overtopping the junipers, move the community on from brush to woodland. White birch, a short-lived tree vulnerable to storm damage and the incursions of fungi, will like the pines be replaced in time by shade-tolerant hemlock and hardwood species.

Whatever the route, profound changes occur in the animal life within the community in the transition from shrub to forest. In the Catskills, the most abundant birds of the beech-maple-hemlock stage are the red-eyed vireo, hermit thrush, veery, and four warblers, the blackthroated green, Blackburnian, blackthroated blue, and the ovenbird, none of which were prominent before. The shrub stage species are scarce or entirely missing in this forest. Although food webs continue to broaden and become more complex, with many more reducers in the ecosystem, there is generally less browse at ground level and deer populations decrease.

Obviously we have a problem here: to maintain maximum diversity, to keep within our environment as many kinds of living things as possible, we must maintain a landscape with areas in a variety of stages of succession. Old field species do not thrive in climax forests and *vice versa*. Managers of natural preserves, including our National Parks and Forests, are becoming increasingly — though often unwillingly — aware of this, realizing that in many instances fire and storm damage are inherent and necessary parts of the ecosystems they are trying to protect. The effects of man inadvertently increasing the diversity of successional stages may be seen at many levels; one of the more pleasant instances involves the invasion of brush and forest edge species into New England and upstate New York from the south in recent years. The cardinal, mockingbird, and tufted titmouse were rarely seen here before the 1950's. Now these attractive birds are fairly common and, while several theories (including climatic change and an increased number of people feeding birds in winter) have been brought forward to explain their presence, the most likely explanation is that man has increased the amount of habitat these species favor and thus allowed them to move north.

There is something comforting in the concept of succession to many people, the reassurance that natural systems, though disturbed, possess the means to right themselves and that human activities, disruptive as they might seem, may not have had as disastrous effects as one might otherwise suppose. Some damage is, however, irreversible — the loss of a species through extinction, for example — while recovery from other types of disturbance may be so slow that for all intents and purposes the harm is beyond repair. Certain ecosystems are particularly fragile (the alpine zone on mountains, certain coastal areas), and deserve the most careful treatment if they are to be maintained. Thoreau concluded his essay on the succession of forest trees with a discussion of the vitality of seeds, stating "Though I do not believe that a plant will spring up where no seed has been, I have great faith in a seed. . . . Convince me that you have a seed there, and I am prepared to expect wonders. I shall even believe the millenium is at hand. . . ." His last sentences strike a more pessimistic note, a reflection that mankind in general prefers deception to true values and "darkness rather than light." One might hope that among the benefits of the current revival of interest in the natural environment and its ecology is a realization of our dependence on those natural processes of which Thoreau was such a skilled observer and proponent.

7.

Natural Cycles

Several times in this book we have referred or will refer to the concept of cycling nutrients and other chemical compounds through the biosphere. We discussed the movement of energy and organic matter from one level of a food chain to another in Chapter 4. In Chapter 11 we will describe the cycling of nitrogen, carbon, and phosphorus by earthworms in the soil. In Chapter 17 we will explain how an excess of nitrogen, phosphorus, or potassium in an aquatic ecosystem can accelerate eutrophication, while in Chapter 19 we will note the importance of the soil microbes in breaking down chemical compounds, thus enacting a natural recycling process. Natural cycles are, indeed, an important aspect of ecology. They will be dealt with in detail in the next several chapters.

Important elements for the growth of living organisms include: carbon, hydrogen, oxygen, phosphorus, potassium, nitrogen, sulphur, calcium, iron, magnesium, boron, zinc, chlorine, iodine, and fluorine. These materials flow from the non-living to the living and back to the non-living again in a circular path which scientists refer to as a "biogeochemical cycle." This term, *biogeochemical*, can be easily understood if we break it down into its component parts: *bio* refers to the living organisms in the cycle; *geo* refers to the water, rock, and soil elements of a cycle; and *chemical* refers to the chemical processes involved in any cycle.

During the last two decades much work has been undertaken in a branch of ecology known as *systems ecology*. In the systems approach, scientists use a "macroscope" to view an ecosystem and try to generalize and, if possible, eliminate details. Two major goals of this approach are: 1) to study exchanges made across boundaries into and out of ecosystems, and 2) to compartmentalize different parts of an ecosystem. Three compartments are associated with mineral cycling in an ecosystem: the living organisms, dead organic detritus (i.e., remains of dead plants and animals), and available inorganic minerals.

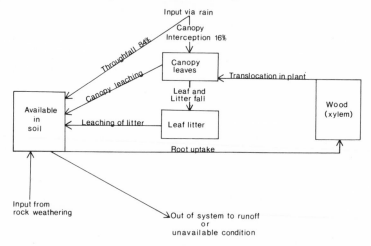

A compartment model of the cycling of elements in a forest ecosystem.

An example of a compartment diagram is shown in the accompanying figure, which depicts the cycling of elements in a forest ecosystem. Elements enter this system in two ways: they are either carried into the system via raindrops or are released to the system as rocks become weathered.

During a rainstorm, some chemicals, about 16% of those carried in precipitation, may adhere to leaves in the forest overstory or canopy, but most chemicals, about 84%, will fall through the leaves (through-fall) and will become available to plant roots in the soil. But, even those elements which are intercepted will eventually be transported to the forest floor, either as leaves drop to the ground or as compounds are washed (or *leached*) out of the canopy. Some chemicals which adhere to leaves during one rainstorm may be washed down a tree trunk during a subsequent storm. When these chemicals reach the ground they may be stored in the accumulation of dead leaves and conifer needles. Additional rain washing through the leaf litter will leach elements out into the soil below. Elements in solution in the soil become available for uptake by roots of forest trees, shrubs, and herbs. After the chemicals are inside the plant roots, they may be carried upward through a group of cells called *xylem*. Some nutrients may be stored for decades in the wood of forest trees or carried from one part of the plant to another through the process of *translocation*. Other elements may be retained in the leaves to begin the cycle over again.

Five of the most important elements for the growth of all living organisms are carbon, hydrogen, oxygen, nitrogen, and phosphorus. In the next few pages we will describe how each of these essential chemicals is cycled through the biosphere.

We will begin with the water cycle. A single molecule of water (H_2O) contains two of the five elements; in a unique chemical bonding, two atoms of hydrogen join with one atom of oxygen to form one of the most important compounds for all living creatures.

A process, fundamental to the cycling of water, is *transpiration*, or the loss of water vapor from the surface of leaves. As much as 98% of the total amount of water absorbed by the roots of plants escapes by transpiration (see accompanying figure).

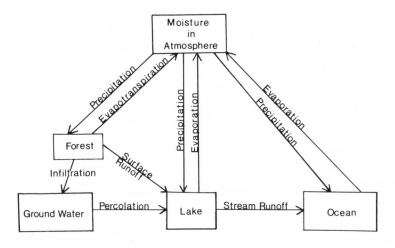

The water cycle.

Another important and perhaps more generally understood process in the cycling of water is *evaporation*. Here light energy is absorbed and used to perform the work of changing water from a liquid to a gaseous form and lifting it into the atmosphere.

A handy equation, which is useful to ecologists, climatologists, and foresters alike, is: Rainfall (RF) = Runoff (RO) + Evapotranspiration (ET). In this formula, evaporation and transpiration are added together for convenience. If any individual knows any two of these values, then the value can be readily calculated. Runoff is the most variable of the three hydrologic parameters, precipitation is less variable than runoff, and evapotranspiration is less variable than precipitation.

In Massachusetts an average of about 50% of annual precipitation goes to stream flow, compared with 60% in the White Mountain region to the north and 40% in the Piedmont region to the south. Generally in New England, precipitation is almost equally distributed throughout the year. However, in spite of the even precipitation, stream flow is not equally distributed. Snow storage on the ground in winter, snow melt in spring, and evapotranspiration in summer cause inequities in stream flow.

Soil characteristics play an important role in the water cycle. *Infiltration* (or the rate at which water will be absorbed by a soil), water storage capacity, and percolation influence the amount of water flowing over the soil surface, the amount retained in the soil, and the timing of water movement through the soil. These in turn influence the amount and timing of water that reaches the streams.

A second cycle, essential to life as we know it, is the oxygen cycle. Here, two highly important processes come into play — photosynthesis and respiration. During photosynthesis, a process carried out by all green plants, water and carbon dioxide are converted in the presence of light and the green pigment *chlorophyll* to oxygen and food. The oxygen is lost from the leaf as a byproduct while the food molecules are transported for use elsewhere in the plant. In the potato, for instance, much of the food produced in the leaves will be translocated for storage in a fleshy tuber. That tuber may in turn become a food source for a herbivore. Food and atmospheric oxygen are utilized by all living organisms during the energy-releasing process of respiration. At the end of that process carbon dioxide and water are released as waste products — and the cycle begins again.

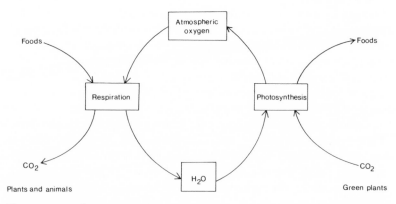

The oxygen cycle.

It has been estimated that oxygen accounts for 21% of the weight of the atmosphere, or weighs roughly 1.1×10^{21} grams. Much more oxygen is bound up in water molecules, mineral oxides, and salts in the earth's rocky crust, but this pool is not available to the ecosystem.

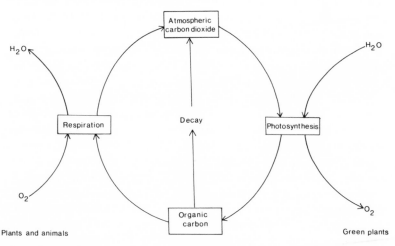

The carbon cycle.

Again, in the carbon cycle, photosynthesis and respiration are complementary. Photosynthesis assimilates carbon dioxide (CO_2) entirely into carbohydrate, while respiration converts all the carbon in organic compounds into carbon dioxide. Plants assimilate roughly 105 × 10^{15} grams of carbon each year, of which about 32×10^{15} grams are returned to the carbon dioxide pool by plant respiration. Some carbon dioxide is released to the atmosphere as dead plant and animal remains decay. The combustion of coal and oil also adds carbon dioxide to the atmosphere yearly, so much so that the carbon dioxide content of the air we breathe has risen measurably during this century. Various analyses of the world circulation of carbon suggest that human activities in the near future could release large additional amounts of carbon dioxide into the atmosphere with results that are substantially unpredictable, as we will discuss in Chapter 20 on air pollution.

Obviously, the oxygen, carbon, and water cycles are intricately interrelated. Green plants are instrumental to all three cycles. By taking water in through their roots, using some water to carry on the process of photosynthesis, and losing some into the atmosphere, plants affect the water cycle. During photosynthesis, green plants produce carbon compounds and release atmospheric oxygen, both of which are utilized by all organisms in carrying out respiration. The uptake of water by plant roots is also important for the acquisition of vital minerals and other essential nutrients; a liquid medium is likewise essential for the transport and later incorporation of chemicals into plant tissue during growth.

Another important, but very complex, cycle is the nitrogen cycle (see accompanying diagram). Land animals, including humans, live in a sea of air that is 79% nitrogen. However, the large pool of gaseous nitrogen in the atmosphere, estimated at around 3.85×10^{21} grams, cannot be used by most organisms. As it exists in the atmosphere, nitrogen is an inert gas except to the relatively few creatures that have the ability to "fix nitrogen" or incorporate it into a chemical compound that can be utilized by plants and animals. Nitrogen is essential to all living things because it is found in protein molecules which are composed of nitrogen-containing amino acids and in nucleic acids, which are the building blocks of chromosomes. These in turn provide structure for cells and regulate all biological functions.

The largest single natural source of fixed nitrogen is undoubtedly terrestrial microorganisms and associations between such microorganisms and plants. Nitrogen is "brought to earth" primarily by the nodule bacteria on the roots of legumes and other plants, although some may also be fixed by lightning during thunderstorms. Root nodule bacteria infect the roots of their host plants and produce nodules or tubercles which may eventually house millions of the bacterial cells. These nodule bacteria live in a symbiotic relationship with their hosts, and over millions of years have evolved unique partnerships with such diverse green plants as alfalfa, peas, alders, and ginkgos. However, not all nitrogen is fixed by nodule producers; some is also fixed by nonsymbiotic bacteria, which live free in the soil.

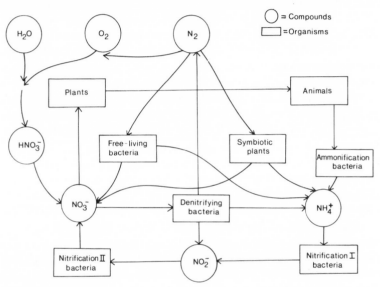

The nitrogen cycle.

Another important group of soil microbes are the *ammonification bacteria*. "Ammonification" refers to the process by which the nitrogen of organic compounds (primarily amino acids) is converted to ammonium ion $(+NH_4)$. This process occurs when microorganisms decompose the remains of dead plants and animals. Another step in the cycle takes place when microorganisms of the genus *Nitrosomonas* convert ammonium ions to nitrite ions $(-NO_2)$ in the presence of oxygen. *Nitrosomonas* belongs to the group of microorganisms called nitrifying bacteria; their sole source of energy comes from the conversion of ammonium ion to nitrite ion. In other words, they are able to obtain energy from inorganic compounds.

Another specialized group of microorganisms in the genus *Nitrobacter* is able to transform nitrite ions to nitrate ions ($-NO_3$), and thus plays a critical role in the cycling of nitrogen. Plants are readily able to absorb nitrates through their roots, and eventually convert them to amino acids and amino acids into protein. The plant tissue may then be eaten by an animal, and atoms of nitrogen will become part of animal protein. Eventually, when the animal dies, ammonification bacteria invade its body, converting the amino acids to ammonium ions, and the cycle begins again.

One final group of microbes plays a critical role in the nitrogen cycle. In the soil there are numerous kinds of denitrifying bacteria that, if living in the absence of oxygen, are able to use either the nitrite or nitrate ion as a source of energy. During the process of denitrification, these soil bacteria may release some nitrogen into the atmosphere, where it will exist as an inert gas until such time as the nitrogen fixers return it back to the soil.

The last cycle we will discuss is the phosphorus cycle. Organisms require phosphorus at a high level as a major constituent of nucleic acids, cell membranes, energy transfer systems, bones, and teeth. But even the "high level" required for phosphorus only amounts to about one-tenth the level required for nitrogen. A critical difference between the phosphorus cycle and those described earlier is that there is no phosphorus available in the air. Therefore, the phosphorus cycle involves only the soil and the aquatic components of an ecosystem. Here the decomposers (bacteria and fungi) play crucial roles in releasing phosphorus from animal tissue and feces, and from dead plant material.

Plants assimilate phosphorus as phosphate ion ($-PO_4$) directly from soil and water (see diagram). Many kinds of rocks contain phosphorus. When such rocks are eroded by water, minute amounts of phosphorus dissolve, becoming available to plants, and thus entering the biogeochemical cycle. But, acidity can affect the availability of phosphorus to plants. Under acid conditions, phosphate is converted to highly soluble phosphoric acid (H_3PO_4), and may be leached out of the system. Also, a certain amount of phosphate runs off the land each year, eventually finding its way to the oceans, where most of it is lost to terrestrial ecosystems.

Phosphorus is important in all ecosystems, but its presence is particularly significant in aquatic communities. In 1840, Justus

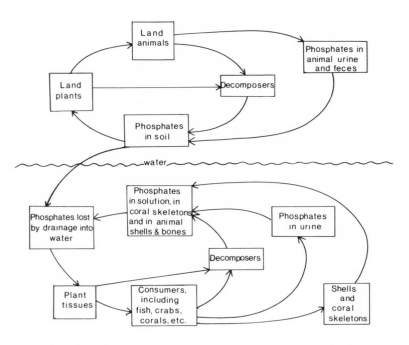

The phosphorus cycle.

Liebig made the observation that "the growth of a plant is dependent on the amount of foodstuff which is presented to it in minimum quantity." Phosphorus is required by plants in such quantities that it is often a limiting factor in plant growth because of low availability. When phosphorus is the limiting factor in the growth of freshwater algae in a lake or slow-moving stream, phosphate-containing detergents added to the water in the outfall of sewage systems may produce a spectacular bloom of algae, the additional nutrients stimulating an increase in the rate of algal reproduction. This period of rapid growth continues until the available supply of another essential element — perhaps nitrogen — is taken up by the algae. The process of cultural eutrophication (to be discussed in Chapter 17), where a critical nutrient such as phosphorus is added to a system through human-related activities such as sewage outfall or agricultural runoff, is a clear example of Liebig's law in operation.

8.

Natural Cycles in Operation

We will now turn our attention to a series of studies undertaken to learn more about how ecosystems in the White Mountains of New Hampshire function. These studies were specifically designed to investigate the energy and biogeochemical relationships of a northern hardwood forest located in the Hubbard Brook Experimental Forest in West Thornton, New Hampshire. The facility there, maintained and operated by the U.S. Forest Service, was established in 1955 and is the major laboratory for research pertaining to management of forested watersheds in New England. The Hubbard Brook site lends itself well to an ecosystem study, since discrete watersheds can be readily defined and compared with one another.

The Hubbard Brook Experimental Forest has a rugged terrain and is covered by an unbroken woodland of northern hardwoods, with spruce and fir present at higher elevations. The climate is typified by short, cool summers, with the average temperature in July about 19° Centigrade, and long, cold winters, with the average temperature in January averaging −9° Centigrade. Six small watersheds were chosen for biogeochemical studies in the mid-1960's; each watershed is entirely forested by uneven-aged, second-growth northern hardwoods, primarily sugar maple, beech, and yellow birch, mixed with a few red spruces and balsam firs.

During the late fall and winter of 1965, all vegetation on one watershed was cut and subsequently treated with herbicides in an experiment designed to determine the effect on: (1) the quantity of stream water flowing out of the watershed, and (2) the fundamental chemical relationships within the forest, including nutrient relationships and eutrophication of stream water. Precipitation falling into the six watersheds was measured with a network of eleven rain gauges, while runoff was measured continuously throughout the year at stream-gauging stations (see photograph). These two measurements provided information on the hydrology of each watershed, including evapotranspiration and the input-output data for individual storms. In order to determine nutrient concentrations in various parts of the watershed, chemical analyses were performed on precipitation and stream samples.

Stream gauges, such as this one installed by the United States Geological Survey, measure runoff from watersheds. / M. Holland

Characteristics of these watersheds have now been monitored for over two decades. The annual distribution of precipitation and runoff for Hubbard Brook Forest is fairly uniform throughout the year, and is typical of moist, temperate locations.

Records kept by the Forest Service between 1955 and 1963 indicate that the average annual rainfall into the sytem was 123 centimeters, while average annual runoff was 72 centimeters. Thus, an average value of 51 centimeters can be calculated for evapotranspiration. The impact of the clearcutting operation on the water cycle was to increase runoff, reduce evapotranspiration by 70%, and increase the velocity of stream flow. Since 1969, the clearcut watershed has been virtually undisturbed and left to recover. The vegetation has been allowed to regenerate, and the watershed has undergone succession. As one might expect, values for stream flow and evapotranspiration in the manipulated watershed are now very similar to the values for undisturbed watersheds — runoff has decreased and evapotranspiration has increased. Here, the role of forest vegetation in the hydrologic cycle is dramatically illustrated.

Results from studies of nutrient cycling at Hubbard Brook are even more dramatic. The natural concentration of nitrate in stream water from undisturbed forests shows a seasonal cycle, being higher from November to April than it is from May to October. The decline in May and low concentration throughout the summer are correlated with a heavy demand for nutrients by the vegetation, as well as with the generally increased biological activity associated with a warming of the soil. Beginning in June, 1966, the concentration of nitrate in the deforested watershed rose sharply, indicating that nitrate was rapidly flushed from the system in drainage water. However, by 1971 nitrate concentrations in stream samples from forested and clearcut systems were the same. During a mere two years with no disturbance, the forest had recovered sufficiently to begin once again to accumulate and cycle large amounts of nitrogen.

Research on a forested, undisturbed watershed at Hubbard Brook has given scientists broader insights into the annual nitrogen budget for northern hardwood forests. Each year 68% of the nitrogen added to the ecosystem comes from nitrogen fixation, while 32% of the amount entering the system comes from precipitation. Moreover, of the total nitrogen entering a watershed during a year, about 81% is held within the ecosystem. Further studies suggest that storage by vegetation of the critical elements nitrogen and phosphorus greatly exceeds stream output; therefore, the role of forest vegetation and soil in preventing nutrient loss is most important.

One of the initial goals of the clearcutting experiment was to determine the relation between removal of vegetation and stream eutrophication. The deforestation experiment resulted in severe pollution of the drainage stream from the ecosystem. From 1966 until 1970, the nitrate concentration in stream water exceeded almost continuously the maximum concentration recommended for drinking water. As a result of the increased temperature, light, and nutrient concentrations, and in sharp contrast to the undisturbed watersheds, a dense bloom of algae appeared in the stream draining the deforested watershed during those four summers.

Important conclusions can be drawn from these studies. Five points emphasized by the Hubbard Brook research team summarize recurrent themes in this book. First, an ecosystem is a highly complex natural unit composed of organisms and their inorganic environment. Second, these parts of an ecosystem are intimately linked by the natural biological, geological, and chemical processes that are part of the ecosystem. Third, the uptake of water and the storage of nutrients by vegetation is critical to maintaining a healthy balance within an ecosystem. Fourth, the stability of an ecosystem is linked to the orderly flow of nutrients between the living and the non-living components of the system. Fifth, individual ecosystems are linked to their surrounding land and water ecosystems, as well as to the biosphere in general.

9.

Disruption of Natural Cycles

So far in discussing cycles we have talked about the natural cycling of five elements in the biosphere and described some of the knowledge gained by conscious manipulation of a small New England watershed. Now we will focus on how people have unwittingly disrupted biogeochemical pathways. One of the ways in which mankind has unintentionally disrupted natural cycles has been through the phenomenon of acid rain. A product of the industrial boom that followed World War II and the volumes of air pollutants it has generated, *acid rain* is, quite simply, rain water which has become acidified. Atmospheric pollutants, mainly sulfur and nitrogen dioxides, combine with water in a chemical reaction induced by sunlight to produce sulfuric and nitric acid. Thus, acid rain impacts the cycling of at least three substances — water, nitrogen, and sulfur.

The most thorough studies to date on the effects of acid precipitation on lakes and streams have been done in Scandinavia. A marked increase in the acidity of thousands of lakes and rivers in southern Norway and Sweden during the past two decades has been linked to acid from precipitation. In turn, this high acidity has caused the decline of several species of fish, particularly trout and salmon. Losses of trout in Norwegian lakes have been observed since the 1920's. However, the number of lakes without trout has grown significantly during the past fifteen years, coinciding with a substantial rise in the combustion of fossil fuels in industrial and urban regions in Europe. Newly hatched fish are most vulnerable to acid, and the death of eggs and young fish is probably the primary reason for the disappearance of Norwegian fish populations. This problem points out an important ecological principle: organisms may vary in their sensitivity to environmental degradation at different stages of their life cycles. Thus, an organism such as a trout will survive in a particular habitat such as a brook only if every stage in its life cycle is tolerant of any pollutants present.

In eastern North America chemical analyses of particulate matter in acid rain have revealed that industrial components such as fly ash and soot are present even in such isolated collection points as the Adirondacks. Studies by a fishery biologist indicate that 52% of the 214 high-elevation lakes in the Adirondack State Park were highly acidic, and that nearly 90% of these acidic lakes were devoid of fish life.

Changes in the chemistry of aquatic environments have had impacts on other creatures besides fish. As lake acidity increases, decreases in numbers and activity of bacteria and protozoans have been observed. These organisms serve an important role as decomposers, breaking down dead plant and animal remains and releasing their nutrients for use by the producer organisms. Leaf decomposition is diminished in acidified lakes, and with an overall slowdown of decomposition there has been a decrease in the cycling of nutrients, particularly phosphorus. Thus, acid rain has an impact on the cycling of a fourth substance, phosphorus.

The effect of acid rain on different bodies of water can be highly variable, depending on the soils and geology of an area. Rain water, after falling on the ground, flows over and through leaf litter, soils, and bedrock as it moves slowly toward the sea. During this journey, acids in the rain water have a chance to react with and be neutralized by basic salts. These salts are known as buffers because they act to neutralize any acid. Their volume determines the buffering capacity of a soil or water body. The greater the buffering capacity, the greater the resistance to the impacts of acidification. In the Hubbard Brook ecosystem, neutralization of acid rain is rapidly and largely (about 75%) accomplished in the upper soil layer through bases leached from various biologic materials. Streams throughout New Hampshire and Vermont tend to become less acid as water moves further downstream and is buffered by basic salts from streambank soils.

Rainfall in the predominantly rural Connecticut River valley is also acidic. However, some recent studies provide a note of encouragement. Research in the area on salamanders, an organism adversely affected by acid rain, along with results from work at the Pleasant Valley Wildlife Sanctuary in Lenox, Massachusetts, indicate numerous populations of spotted salamanders are surviving the onslaught of acid rain. These findings offer the possibility of repopulating ponds where salamanders have disappeared.

Is there some hope for the survival of other acid-tolerant animals? Researchers hope to develop acid-resistant stocks of fishes. Still, the only long-term solution to this problem appears to be adoption of and strict enforcement of stringent air pollution emission standards, not only in the Northeast but throughout the whole world.

10.

Managing Natural Cycles

In this chapter we will focus on ways in which researchers and managers have been able to work with natural processes to develop sensible management strategies for wise use of our natural resources. Since trees appeared on the evolutionary scene, forests and stream flow have been interrelated. Although this relationship has not always been completely understood, by the turn of the present century most scientists agreed that forests exert a beneficial influence on stream flow. Forests, they observed, provide a protective cover for the soil surface, allowing rainfall to infiltrate into the soil rather than flow over the surface, eroding the soil away.

Increasing population, improved technology, and subsequent demands for multiple land use have placed new values on forest and water resources, particularly in highly urbanized areas. For example, in the densely populated Northeast the primary source of most surface water is stream flow from forested watersheds. In this region alone more than two million acres of land are under the control of municipalities, private water companies, and state and federal agencies for management as either water source areas or protected lands for municipal water supplies. Continuing urban sprawl has increased water demands, making it imperative that these watersheds be managed for maximum water production and high water quality. At the same time there are sometimes conflicting public pressures to use these areas for recreational purposes, wildlife habitats, and harvesting of timber products.

During the past thirty years, forest watershed management

research has produced much useful information about water yield augmentation and water quality protection. Data collected over two decades exists, for example, from the 36,000 acre Pequannock watershed in New Jersey. The Pequannock, owned by the city of Newark, provides about one-half of the water supply of Newark and adjacent cities. In 1959, the Newark Division of Water Supply and the Northeastern Forest Experiment Station initiated a cooperative study of the effects of chemical control of vegetation on water quality and yield. Although application of herbicides on watershed lands has been generally discouraged during the last ten years, the results of the Pequannock venture are still instructive.

Partly because of their substantial transpiring surfaces and extensive root systems, trees use far more water than other forms of vegetation. Nonetheless, there are differences in transpiration rates among different types of trees. Stream flow from a hardwood-forested watershed is likely to be two or three inches a year greater than from a conifer-covered watershed because of the greater interception of snow and rain in conifers during the dormant season. Conifers may also lose more water by transpiration than hardwoods, especially in early spring and late autumn.

In the fall of 1965, all large trees in one watershed of the Pequannock forest were injected with herbicide. By the summer of 1966, live overstory vegetation had been reduced to 31% of the watershed area. Grasses and herbs increased in response to the reduction in overhead shading and in 1966 covered 26% of the watershed. Reduction in vegetative cover following the injection of herbicides resulted in a water yield increase of 4.67 inches, which translated into an increase in water production of 127,000 gallons of water per acre per year — a substantial increase in water yield. (If, just for the sake of discussion, this treatment were applied to the entire watershed, a practice which we are sure our readers realize would be foolhardy — 36,000 acres × 127,000 gallons — the Pequannock watershed would yield a total increase of 4.5 billion gallons of water per year.)

One may very well ask at this point, "What impact do experiments such as the one in the Pequannock forest have on water quality?" And, a second question naturally follows: "Can a watershed manager, faced with the prospect of allowing multiple-use activities such as logging and recreation in a watershed, produce a quality of water that will meet high drinking water standards?" Research designed to answer these questions is presently underway; preliminary indications are that at least some timber production and harvesting practices are compatible with high water quality production.

In several experimental watersheds logging operators were given freedom to harvest and transport timber in whatever manner they wished. Roads, skid roads, and trails were placed with little or no regard to water quality. Careless logging was permitted, for example, in the Fernow Experimental Watershed in the central Appalachians, where soils are primarily silt loam. In areas where forests were clearcut, the turbidity of the streams increased greatly. During the same period on a nearby undisturbed area the streams remained clear. In all experiments in the Fernow Watershed, the major damage to water quality occurred during and immediately after logging. Once logging ceased, the exposed mineral soil quickly was covered by a protective layer of grasses and herbs.

At the Hubbard Brook Experimental Forest, after a decade and a half of studying the effects of clearcutting, researchers concluded that for it to be an ecologically acceptable practice for harvesting timber in the northeastern United States, it must be coupled with carefully designed safeguards. Recommendations included the following: (1) cutting should be limited to sites with relatively fertile soils on modest slopes, (2) clearcuts should be small enough to minimize excessive runoff of dissolved nutrients and eroded material, (3) the cutting and harvesting procedure should do little harm to the forest floor, (4) roads and skid trails should cover a minimal area, and (5) sufficient time should be allowed before the next cut of the forest for it to regain by natural processes the same amounts of nutrients and organic matter lost as a result of timber harvest. One logging technique which has been historically successful in preventing damage to soil is removal of logs in winter over a protective snow cover.

Watershed management is extensively practiced at many reservoirs including the giant Quabbin Reservoir, built in the 1930's by the Massachusetts Metropolitan District Commission, a state agency which supplies water to 34 metropolitan Boston and ten central Massachusetts communities. Approximately 65 miles west of Boston, this reservoir covers more than 39 square miles of river valley and is surrounded by a forested, hilly watershed. Lying in the valley of the Swift River, a tributary of the Connecticut, Quabbin was created by the construction of two large earthen dams (see photograph) that impound a storage volume of 412 billion gallons. The reservoir is edged by approximately 165 miles of scenic shoreline and contains numerous rocky and mountainous islands.

The Winsor Dam at the southern end of Quabbin Reservoir is one of the largest dams in the eastern United States. / M. Holland

Since the Swift River is part of the Connecticut River watershed, its damming during construction of Quabbin constituted an indirect taking of water from the Connecticut River for Boston's water supply. Construction of Quabbin also required relocating boundaries of six towns and three counties and eliminating from corporate existence the towns of Enfield, Dana, Greenwich, and Prescott. About 2,500 people living in 650 houses were forced to move, causing much inconvenience and disgruntlement.

The forests at Quabbin consist of red, black, scarlet, and white oak, red and white pine, red maple, birch, hemlock, ash, and spruce. Some of the pine and spruce, planted 33 to 44 years ago, are very dense. Conifers have been traditionally favored for reservoir shoreline vegetation, since hardwood leaves in the immediate vicinity of the reservoir discolor water and clog intake pipes. Moreover, conifers reduce wave action (particularly on small reservoirs) by lowering wind velocities in winter, preventing churned up bottom sediments from forming turbulence. However, since conifers lose more water than hardwoods by transpiration, a change from conifers to deciduous species can significantly increase the amount of water available.

As early as December, 1957, the importance and desirability of carrying out watershed research in Massachusettts was acknowledged. The 403-acre Cadwell Creek watershed, a part of the drainage basin of Quabbin Reservoir, was chosen as the focal point of a study and in June of 1961 two gauging stations were established in anticipation of a watershed treatment program for the area.

During the summer of 1967, 49 acres of forest along the edges of the two major streams of the watershed were treated with herbicides. The following winter, an additional area of roughly 56

Cadwell Creek drains into Quabbin Reservoir on the western side, and has been the focal point of management studies for over a decade. / M. Holland

acres of forest overstory was logged. Results of this study suggest that annual increases in water yield on the order of 20% could be obtained by an integrated management system which removed only one-third of the original forest cover in the small sub-watershed.

Hunting on the Quabbin Reservation is strictly prohibited. The heavy deer population, browsing on the hardwood sprout growth which shot up following the herbicide treatments, keeps the clearcut areas in an open, meadow-like condition. Indeed, the deer seem much more successful in controlling regrowth than herbicides.

Equally important in determining the character of the watershed is the beaver. Distributed throughout the watershed are approximately 1,200 acres of ponds and marshes created solely by hard-working beavers; these areas supply homes and food for nearly every bird, animal, or fish in the area. During timber harvesting overmature, diseased, severely defective, and dying trees are cut to maintain a desirable and high-quality stock. Thus, the lower forest densities required by watershed management may improve both tree quality and habitat diversity for wildlife.

Fishing in Quabbin is reputedly excellent and generally falls into two broad categories, coldwater and warmwater, which reflect the requirements of two different groups of fish. The salmonids, such as lake trout, rainbow and brown trout, and landlocked salmon, along with their principal forage, smelt, are examples of coldwater species. The various species of bass, pickerel, white and yellow perch, and bullheads are considered warmwater fish. To protect the high water quality and to maintain the fishing at Quabbin, forested areas are cut only in accordance with the strictest conservation principles, and thus erosion and nutrient runoff are minimized.

A sound management program has been implemented at Quabbin to integrate the following four goals:

1. To create and maintain an additional water yield of 10% annually from the Quabbin Reservation,
2. To improve the health and quality of the Quabbin forest,
3. To maintain healthy populations of native wildlife, and

4. To maintain, improve, and protect an aesthetically pleasing landscape.

Over the years, the Reservoir staff, in addition to timber sales, has sold wood for pulp and for guard rail posts. In the last few years the demand for firewood has increased dramatically, and in 1978 the forestry staff sold over $20,000 worth of cordwood. Thus the Metropolitan District Commission, through implementation of an integrated watershed management program using sound scientific principles and strong conservation techniques, has been able to insure preservation of timber, a renewable resource, and increased yields of water, a recyclable resource, while realizing profits on the sale of each.

Over millions of years, pathways have evolved which cycle the finite resources of the biosphere. For centuries the cycling and recycling of water has been recognized, as the perceptive author of *Ecclesiastes* noted: "All the rivers run into the sea; yet the sea is not full; unto the place from whence the rivers come, thither they return again." At last we are beginning to realize the importance of wise resource management. Ultimately, the best management strategy involves a strong conservation program which decreases demand and stresses recycling of our natural resources.

For those who would like more information on natural cycles, we recommend the following references:

Bormann, F. H., and G. E. Likens, "The Nutrient Cycles of an Ecosystem," *Scientific American*, October, 1970.

Farb, Peter, *Living Earth* (Pyramid Publications, Inc., New York, 1959).

Ludlum, David, *The Country Journal New England Weather Book* (Houghton Mifflin Company, Boston, 1976).

Storer, John, *The Web of Life* (The New American Library of World Literature, Inc., New York, 1953).

11.

Soil and Soil Conservation

All of us interact with soil, either through the gardens we tend or the trails we hike. We may call it "dirt" or "mud" or "earth," but how many of us really understand what is meant by soil structure? A good soil is generally defined as one that consists of five components: (a) the rock particles that are its foundation, (b) the organic matter given it by dead plants and animals, (c) a community of living plant and animal organisms, (d) air, and (e) water.

Soils begin with the weathering of rock. Exposed to the combined action of wind, water, and temperature, rock surfaces peel and flake away. Water seeps into crevices, freezes, expands, and cracks the rock into smaller pieces. Rocks are broken apart by the expansion of roots in crevices and are also dissolved by acids secreted from the roots. Eventually, the rock is broken down into loose material; it may remain in place, but more often much of it is lifted, sorted, and carried away. Material transported from one area to another by wind is known as *loess*, that transported by water as *alluvium*, and by glacial ice as *till*. In a few places soil comes from accumulated organic matter such as peat.

Alluvial deposits have formed much of the state of Mississippi. Soils deposited by river systems are among the richest on the planet, and a large percentage of the world's population is fed from those formed by the Mississippi, Yangtze, Ganges, Amazon, and Nile. The high fertility of these soils is due to the fact that the particles were drawn from diverse areas, with a whole range of rocks and minerals to provide abundant quantities of every nutrient essential for plant growth. Alluvial soil fertility accounts for the high agricultural productivity in the Connecticut River valley. Anyone who has gone for a bicycle ride along the Connecticut River in the summer months is familiar with the acres and acres of onions, potatoes, asparagus, corn, and tobacco growing on the river floodplain in Hadley or Hatfield. (The river and its floodplain will be discussed in greater detail in a later chapter.)

Material transported from one area to another by glacial ice is known as till; a single boulder surrounded by open space is often referred to by geologists as a glacial erratic. / C. J. Burk

Once rock particles have been deposited in a particular location, they are covered by the remains of local plants and animals. This fresh or partly decomposed organic material forms the surface layer, often called the O horizon, which lies above the mineral layer. The surface layer is expanded each fall as herbaceous plants die and as trees such as oaks, maples, and birches shed their leaves. Birds, mice, and insects add their bodies to the organic layer as they die. When a plant dies, a dramatic change takes place in the soil. At that time the dead plant's roots and leaves offer food to the small organisms that are among the most important factors in the soil, the bacteria and the molds. The most significant function of these reducers lies in decomposing the remains of the higher plants and animals, changing them into new chemical combinations that can be used again by succeeding plant generations for food. Such an abundance of food makes the O horizon the region of the soil where life is most abundant. It is subject to the greatest changes in soil temperatures and moisture conditions and contains the most organic carbon.

Members of the soil population do not live in the soil, but rather between the soil grains. The grains make up a framework, varying quite a bit in size. Clay soil, the finest, has particles that are smaller than 1/200ths of an inch in diameter. Particles of sand, the largest, range up to one-twelfth of an inch. Silt is halfway between sand and clay; any particles larger than sand are simply gravel. Each soil grain has a tremendous surface area, and together they offer a variety of habitats for living organisms among the particles.

Beneath the surface of the ground lies another world, with its own chain of life, its predators and prey, its herbivores and carnivores, and its fluctuating populations. The soil is an extremely different environment for life than the one above the surface, yet the essential requirements are the same. Like animals that live elsewhere, soil fauna need living space, oxygen, food, and water. The number of animals that can take advantage of these requirements in the soil is astonishing. In his book entitled *Ecology and Field Biology*, Robert L. Smith lists some of the enormous number of different species found in the soil: 110 species of beetles, 229 species of mites, and 46 species of snails and slugs were found in the soil of a beech woods in Austria; nematodes ranged from 708 million to 81 million per acre in Denmark; earthworms varied from 167,000 per acre in a 300-year-old pasture to 1,450,000 per acre in a beech forest; mites accounted for 83 per cent of all the soil fauna in a pine woods in Tennessee.

Nearly 95% of all insects spend a part of their life cycle below the ground surface. No insect, though, relies more on the soil for incubation than the periodical cicada. Its lifespan of 17 years is spent almost completely in the soil. Little is known about its life underground. When the cicada enters the soil it is a miniscule creature with a light-colored, soft body, only about a twelfth of an inch long. With its sucking beak it feeds itself on the sap of roots. Throughout its many years of life in the ground, it molts about half a dozen times, each time growing a new skin which

The life span of the periodical cicada is spent almost entirely burrowing in the soil. Shortly after the young hatch from their eggs, they crawl into the upper two feet of soil; there they will remain for seventeen years, feeding themselves on juices from the roots of deciduous forest trees.

looks more like the adult. In its last spring as a soil resident, it moves toward the surface. It builds a chamber about six inches below the surface, constructs a chimney leading to the sunlight, and waits, sometimes for weeks, for the final emergence. Toward dusk on emergence day, the cicadas receive some unknown signal; they drill through the ground, and the adult cicadas crawl out. The males then commence their music-making, which to some people sounds more like the whine of a buzz saw. After several weeks of feeding on leaves, mating, and egg-laying, a fungus disease spreads through the cicada population, attacking and killing many individuals. Thus, the adult bodies fall to the ground, adding their carcasses to the O horizon of the forest floor.

Earthworms may have a great influence on the physical structure of the soil. Earthworm activity in the soil consists primarily of burrowing, ingestion and partial breakdown of organic material, and egestion in the form of surface or subsurface casts. Ingested soil is taken during burrow construction, mixed with digestive secretions, and eliminated either as aggregated castings on or near the surface or as a semiliquid in intersoil spaces along the burrow. Casts of soil passed through the digestive tract contain a larger concentration of total nitrogen, organic carbon, and available phosphorus than uningested soil. Surface casting and burrowing slowly overturn the earth. Lower level soil is carried to the top and organic material is pulled down into and incorporated with the subsurface soil to form soil aggregates. In a study of English fields, it was estimated that from four to 36 tons of soil passes through the digestive tracts of the earthworm population living on an acre of land each year. Earthworms provide an important link between the organic matter of the O horizon and the mineral layer below it.

In the past few years people have become increasingly aware of the problems involved when the O horizon is washed away during soil erosion. *Infiltration capacity* is the geologist's term for the rate at which water will be absorbed by a soil. The infiltration capacity of a given soil during a given rainstorm is determined by soil texture, structure, vegetative cover, biologic structures, and conditions of the soil surface. Plant roots, worm boring, animal holes, and other biologic phenomena help increase infiltration capacity. However, any moisture still present from a previous rain tends to lower infiltration capacity. Likewise, extreme dryness, where the soil surface is baked or compacted, lowers the ability of the soil to absorb water. We have all seen heavily travelled trails which suddenly are exposed to a downpour after a couple of weeks of dryness — the path cannot absorb all the water, so part of it is eroded away, leaving a gully through the trail.

In many areas various plants have been introduced to control erosion and conserve the soil. Vegetative cover aids infiltration by preserving loose soil and diffusing the flow of water, increasing infiltration opportunity. In one experiment a planting of winter rye grass was grown for four months in a box with less than two cubic feet of earth. In that time the root system had developed 378 miles of roots and an additional 6,000 miles of root hairs. In nature these growing roots bind together the rock particles that form the soil, and thus prevent erosion.

Contour plowing is one successful conservation technique highly recommended by the Soil Conservation Service for farming in hilly regions. Here this method is practiced on a farm in Pennsylvania. / Courtesy of the U.S. Soil Conservation Service

In 1935 the Soil Conservation Service was created with the goal of making the country more conscious of soil and moisture conservation. Today this organization is working with thousands of farmers and gardeners across the country in an effort to maintain good soil. One successful conservation technique is to plow furrows along the level contours of a slope, creating thousands of small reservoirs. With this method the water will sink into the soil between the furrows, and the soil will stay in place without eroding. The home gardener knows that seeds planted in a row sloping downhill will wash out with the first substantial rainfall. Thus, planting along hill contours increases moisture retention, and maintains a healthy soil.

Soil structure is indeed complex. As Aldo Leopold said in his conservation classic, *A Sand County Almanac:*

> Land, then, is not merely soil: it is a fountain of energy flowing through a circuit of soils, plants, and animals. Food chains are the living channels which conduct energy upward; death and decay return it to the soil. The circuit is not closed; some energy is dissipated in decay, some is added by absorption from the air, some is stored in soils, peats, and long-lived forests; but it is a sustained circuit, like a slowly augmented revolving fund of life.

Thus, the number of organisms living on the energy in the soil depends on the type of rock particles, the amount of organic matter, and the water infiltration capacity of a given soil. Unfortunately, these complexities are often appreciated only after some part of the system has been disturbed.

12.

Rivers and Floodplains

The system of streams and rivers which now crisscrosses the northeastern countryside is relatively young from a geologist's point of view. One million years ago, at the start of the ice age, glacial ice from the north advanced southward and covered the Northeast intermittently. The ice sheet melted in southern Connecticut about 14,000 years ago, and later at points northward toward Canada. As the ice melted, a dam of debris was deposited near Middletown, Connecticut. Behind this blockade, meltwater was impounded forming glacial Lake Hitchcock. As the ice front continued to recede, the lake expanded to the north, and water backed up behind the dam for 157 miles north to Lyme, New Hampshire. From examination of the sediment deposited in its waters, geologists know that Lake Hitchcock existed at Springfield, Massachusetts, years before it appeared at South Hadley Falls, and covered the meadows near Middletown, Connecticut, for nearly 6,000 years before its waters reached Northampton.

After Lake Hitchcock had been in existence for a few thousand years, the water level suddenly dropped ninety feet when the blockade at Middletown collapsed. The lake drained away below Charlestown, New Hampshire, and exposed the fine sediment deposits of its floor to the erosional activity of the newly formed Connecticut River. This young river maintained its channel in the lowest part of the lakebed and has since carved prominent terraces in the lake bottom. Today the Connecticut River arises in Canada in the province of Quebec and flows southward for some 400 miles into Long Island Sound, forming the largest river system in New England.

The Connecticut River meanders back and forth in its floodplain. As the river swings from side to side, each loop grows by erosion on the outside and by deposition on the inside. / M. Holland

As we view the river in its present course, we see that the Connecticut tends to wander in flat places, seeking the easiest route in its path to the Atlantic Ocean. These wanderings are called *meanders,* and through time they move progressively downstream. Both erosion and deposition are involved in the formation of a meander. Erosion takes place on the outside of each river bend, where water turbulence is greatest. The material removed from the banks is carried downstream to be deposited in areas of less turbulence. As the river swings from side to side, the meander continues to grow by erosion on the outside of the bends and by deposition on the inside.

From time to time, usually during a flood, the river cuts a new, shorter channel for itself between successive meanders. The abandoned river channel is slowly filled in with silt and debris and sealed at both ends, isolating the old channel loop in the form of a cut-off, or *oxbow lake.* In 1840 the waters of the Connecticut River cut a new channel across the narrow neck of land along the riverbank in Hadley, Massachusetts, straightening the course of the river. Prior to the cut-off the river had flowed an extra three and one-half miles around three hundred acres of land called Hockanum Meadows. The Meadows had belonged to Hadley, but after 1840, the land became part of Northampton, much to the dismay of Hadley residents! Today, the 1840 cut-off is called "The

Oxbow," and the sheltered waters of the abandoned meander are a favorite site for boating and fishing.

Several factors influence the types of organisms which can survive in rapidly flowing waters. Current determines the nature of the bottom sediments, which in turn determine the type of vegetation which can survive on the stream bed. Larger plants modify the physical conditions in a river by collecting silt in their roots and branches, and forming mounds of sediment. Spring rainfall may cause flooding, which increases current speeds and scours away the plants. Rainfall may also increase the turbidity of the water and, hence, cut down the amount of light reaching bottom plants. During normal flow, current speed decreases closer to the river bed and a boundary layer with a thickness of less than an inch is formed. The boundary layer, with its almost non-moving film of water close to the stream bed, is an important site for the attachment and development of young plants.

Today the Connecticut River rises in Quebec and flows southward for some four hundred miles into Long Island Sound, forming the largest river system in New England. Several thousand years ago much of the present Connecticut River Valley was covered by a long, narrow lake known as glacial Lake Hitchcock, formed by the meltwaters of a receding glacier.

Several plant species have evolved unique adaptations to stream life. Plants with strong holdfasts and linear or finely divided leaves often occur in swiftly flowing waters. Mosses, many of which are typical of torrential streams, tend to flatten against the substrate and present a smooth surface to the current. One species of green algae (*Cladophora glomerata*) dominates stream bottoms where the water flows rapidly in much of North America, but never grows in still water. Most organisms adapted to living in active currents cannot survive outside of the currents, since the flow renews the depleted mineral requirements for life and removes the accumulating byproducts of metabolism. For example, the green alga *Oedogonium kurzii* substantially increases its uptake of vital mineral substances in a current, while its uptake decreases in quiet water. This mineral uptake is essential to aquatic producers, enabling them not only to maintain themselves but also to continue to synthesize food for the consumers of the riverine ecosystem.

Many stream inhabitants live in turbulent riffles on the underside of rubble and gravel where they are sheltered from the current. Characteristic riffle insects are the nymphs of mayflies, caddisflies, stoneflies, and dobsons. The larvae of certain species of caddisflies construct cases of sand or small pebbles which shelter them from the full force of the current. Some species have portable houses, the weight of which increases with the velocity of the current. The thickened walls act as ballast to hold the case on the bottom. Other species have cases firmly cemented to the sides and bottoms of stones. Larvae of the water-net caddisfly build a funnel-shaped net, opening upstream and ending in a pebble-bound cemented den of silk. The larvae feed on the minute plants and animals swept into the meshes of the nets, and thus function as primary or secondary consumers in the stream ecosystem.

A streamlined body which offers less resistance to current is typical of many animals of fast streams, such as the black-nosed dace and the brook trout. The black-nosed dace breeds over a gravel bottom in clear brooks from New England to Minnesota. It feeds mainly on small animals such as midge or mayfly larvae and sometimes on fish eggs. The brook trout is found in streams with a maximum temperature of 66° Fahrenheit from the Labrador peninsula to Georgia, but can survive temperatures to 75°

Fahrenheit. The female makes a nest in a riffle over a gravel bed; after the eggs are laid, the male may guard the nest for up to three weeks. Introduced all over the world in suitable waters, the brook trout is considered by many fishermen to be a superior food.

The Connecticut River erodes varved clays at Indian Hill in Hadley, Massachusetts, exposing buried artifacts. / M. Schalk

With the addition of excess silt, the cold, clear trout stream can easily be changed to a warm, murky one, inhabited by fish tolerant of turbid waters and muddy bottom. Siltation, caused by erosion of farmlands, roadsides, construction, and other forms of soil disturbance, is a subtle form of pollution, for it is widespread and often goes unnoticed. Silt settles on the stream bottom, covering sites for insect larvae and other bottom organisms. Caddisflies and mayflies are replaced by bloodworms in such waters. Silty water flowing through the gravel nests of trout causes heavy mortality among the eggs. In this way thousands of miles of trout stream have already been destroyed by siltation inadvertently caused by man.

Naturally, rivers carry some silt downstream each year during the spring flood season. As the snows melt in the mountains up north, the water level rises in the streams and they overflow their banks. When the river floods the surrounding land, it distributes the mud, twigs, and leaves which it has collected upstream. As the water leaves the main stream bed, it spreads out and slows down, depositing its burden of sand and silt along the edge. When the water returns gradually to the main stream bed, new soil is left behind. Floods, then, can be viewed as part of the river's natural cycle for maintaining its nutrient-rich floodplain.

The *floodplain* is a depositional feature of the river valley where sediment is temporarily stored in channel bars, alluvial islands, and oxbows. These geological features are formed as sediment is moved by the current from one place to another along the river channel, but the sediment is eventually held in the floodplain by the roots of the plants growing along the water's edge. Another function of the floodplain is to provide for temporary water storage, to act as a giant sponge.

Studies of floodplain vegetation indicate that certain species are able to withstand periodic flooding and thrive on the rich alluvial sediments. G. E. Nichols studied the Connecticut River floodplain near Windsor, Connecticut, in 1916, and listed as dominant tree species, black willow, sycamore, cottonwood, silver maple, and red ash. In our own research along the Connecticut River floodplain in Northampton, we have found silver maple the dominant species, with pin oak, red ash, black willow, and red maple somewhat less abundant. Without these tree species to naturally control erosion, flood damage downstream would be considerable.

The development of human civilization has been closely linked with the floodplain environment. With the development of mechanical technology, man has acquired increased capacity to control flooding. Natural floodplains have been viewed as useless swamps, and little thought was given to their preservation. Dikes were built to keep rivers in their channels and out of the croplands. Only recently has any planning been devoted to floodplain usage in which man and the river would be compatible. We now know that floodplains may be ideal for many sorts of recreational activities, such as riding, hiking, or camping, but that they are far from suitable sites for housing developments or industrial parks. As we learn more about the delicate relationships between the organisms of the river, the geologic phenomenon of sedimentation, and the communities of floodplain vegetation, we must also learn to use the river's resources while still preserving these ecological features which make those benefits available.

13.

Lakes

Anyone who has spent a summer's night in one of the cabins at the AMC's Three Mile Island Camp can appreciate W. B. Yeats's lines:

> I will arise and go now, for always night and day
> I hear lake water lapping with low sounds by the shore
> While I stand on the roadway, or on the pavements
> grey,
> I hear it in the deep heart's core.

The sound of lake water lapping by the shore is indeed very relaxing, and to many of us who have vacationed at Three Mile Island, it represents an ideal vacation spot and a perfect chance to "get away from it all." But during recent years, Lake Winnipesaukee, along with other lakes of New Hampshire's Lakes Region, has been increasingly exposed to recreational and developmental pressures. More vacation homes are being constructed along the waterfront, more motorboats are churning up the water, and the problems of trash and sewage disposal have grown. We will later look at some of the polluting effects man has had on lakes and streams, but all the possible ramifications of increased development around lakes have not yet been worked out.

Scientists have classified lakes into four categories: oligotrophic, mesotrophic, eutrophic, or dystrophic. *Oligotrophic* waters possess a small supply of nutrients and hence support a small organic production. These lakes are typically deep, continuously oxygenated, and often poor in phosphorus, nitrogen, and calcium.

Oligotrophic lakes are typically deep, continuously oxygenated, and often poor in phosphorus, nitrogen, and calcium. / C. J. Burk

A eutrophic lake is relatively shallow, is nutrient-rich, and often experiences a depletion of oxygen on the lake bottom during the summer. / C. J. Burk

Lakes with a depth and nutrient supply intermediate between oligotrophic and eutrophic are classed as mesotrophic. / C. J. Burk

A *eutrophic* lake has water with a good supply of nutrients and hence a rich production of organic matter. Lakes and ponds of this type are relatively shallow and often experience a depletion of oxygen on the lake bottom during the summer. Lakes with a depth and nutrient supply intermediate between oligotrophic and eutrophic are classed as *mesotrophic*. The *dystrophic* type, associated with bogs, are brown-water lakes, often with an abundance of peat and sphagnum moss and a very high humus content.

As oligotrophic lakes gradually fill in and become shallower, they may develop into mesotrophic lakes. Mesotrophic lakes, through time, may change to form eutrophic lakes, which in turn may become marshes and eventually meadowland. This long-term process is referred to as *natural eutrophication*. Terrence P. Frost described this phenomenon in his article, "The Galloping Ghost of Eutrophy" (*Appalachia* No. 146; June 15, 1968); he wrote, "Left to natural regression, without benefit to man, lakes tend to die over millennial or geological time. It takes eras and eons for a lake basin to degrade to dry land, when it can be truly said that the lake is dead." But, it must be remembered that all lakes do not fill in at the same rate. The speed at which natural eutrophication occurs depends on topography, sediment deposition from neighboring streams, weathering, and degree of fertility.

Probably all northeastern lakes except the artificial ones owe their origin in part to Ice Age glaciation. These lakes have been filling in for approximately 10,000 years. New England's largest and deepest natural lakes include Moosehead, Winnipesaukee, Sebago, and Willoughby. Moosehead Lake, located near the center of Maine, is largest with an area of 331 square kilometers and the third deepest with a maximum depth of 75 meters. Lake Winnipesaukee, New Hampshire's largest lake, with an area of 180 square kilometers is the second largest in New England, but its maximum depth is a mere 52 meters. Sebago Lake in southern Maine is not only third largest, with an area of 116 square kilometers, but also the deepest, with a maximum depth of 96 meters. Lake Willoughby in northeastern Vermont has a very small area in comparison with the others (only 6 square kilometers), yet has a maximum depth of 94 meters. Moosehead, Sebago, and Willoughby are probably all oligotrophic, while some studies have classified Winnipesaukee as mesotrophic.

In 1967 and 1968 P. J. Sawyer and a research team from the University of New Hampshire collected data from several lakes in New Hampshire's Lakes Region. Their findings indicate that Squam, Ossipee, and Winnipesaukee might be classed as mesotrophic, while Lake Winnisquam seemed eutrophic and Newfound Lake oligotrophic. Differences between the lakes are related to steepness of their sides and surrounding land use, as well as to differences in nutrient buildup.

Anyone who has ever gone SCUBA diving during the summer in a relatively deep lake knows the cold sensation that is felt when one hits the *thermocline.* What is a thermocline, and why does it exist? The thermocline is the middle zone of a mid-summer lake where there is a sharp drop in temperature (about 1° Centigrade for each meter of depth). The existence of a thermocline is an indication that the lake is stratified. Such stratification is mainly due to temperature changes from the surface waters down to the lake sediments. The warm, freely circulating surface water, with a small temperature gradient, is the *epilimnion.* At the bottom of the lake is the *hypolimnion,* a deep, cold layer in which the temperature drop is gradual. Between these two is the thermocline. Oxygen stratification during summer nearly parallels that of temperature, with a decrease in quantity occurring with greater depth. The greatest amount of oxygen is generally near the surface, where there is some interchange between water and atmosphere.

Each year the waters of New England lakes undergo seasonal changes in temperature. In autumn, the air temperature falls, and the temperature of the surface drops. Eventually the thermocline sinks, the epilimnion increases to include the entire lake, and the temperature becomes uniform from top to bottom. Even the slightest wind can stimulate lake waters to circulate, and the ensuing fall overturn recharges nutrients and oxygen throughout the lake. As the surface water cools below 4° Centigrade, it becomes lighter, remains on the surface, and freezes. A winter temperature stratification develops where the higher temperature is on the bottom, due in part to heat conducted from bottom mud. When the ice melts in early spring, the surface water, heated by the sun, warms to 4° Centigrade. Aided by strong spring winds, currents mix the water throughout the lake until all the liquid is a uniform 4° Centigrade. This spring overturn stirs bottom nutrients, floating algae, and surface oxygen throughout the lake.

The abundance, distribution, and diversity of lake life are influenced by temperature, oxygen, nutrients, and light. The depth to which producers are found is dependent upon the depth to which light can penetrate. Dominant among photosynthesizers in open water are the phytoplankton, which include diatoms, desmids, and the filamentous green algae. Aquatic life is the richest and most abundant in shallow water, where sunlight can reach the bottom. Common producers along the edges of lakes include submerged plants such as the alga *Chara* and the bladderwort *Utricularia*. *Chara*, relatively common in springs and ponds with a high lime content, reputedly limits development of mosquito larvae. Bladderwort is named for the bladder-like structures near the bases of the leaves. Bladders capture minute water animals, such as water fleas, through the bladder opening and apparently digest them! These submerged plants in turn serve as cover for many of the scavengers of shallow water, including snails or crayfish.

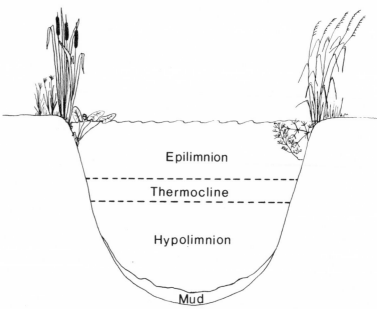

Summer stratification is mainly due to changes in water temperature between surface waters and lake bottom sediments. The epilimnion is the warm, freely circulating surface water, with a small temperature gradient. The thermocline is the middle zone where there is a sharp drop in temperature (about 1°C. for each meter). At the bottom of the lake, the hypolimnion is a deep, cold layer, in which the temperature drop is gradual. (NOTE: in this diagram the horizontal scale is greatly reduced.)

The dominant consumer organisms of open lake waters are fish, such as the lake trout, bullhead, pumpkin seed, yellow perch, and chain pickerel. Found in large, cool, hard-bottomed lakes

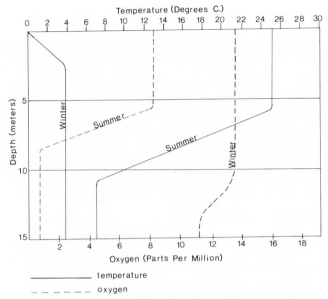

Oxygen stratification during the summer nearly parallels that of temperature, with a decrease in quantity occuring with greater depth. The greatest amount of oxygen is generally near the surface, where there is some interchange between water and atmosphere.

from New England to Montana, the lake trout breeds in water from three to 100 or more feet deep. More common in quiet, mud-bottomed waters are the yellow perch and bullhead. Ranging from Nova Scotia to the Dakotas, the yellow perch feeds on a variety of small animals, including worms, crayfish, and minnows. The bullhead mates in late spring in waters from Maine to Texas and Florida. The male guards the nest of about 2,000 eggs, and after the eggs have hatched, will protect the young until midsummer. Often, conspicuous dark patches may be seen in the water as the male swims about with the young in a school. The chain pickerel and pumpkin seed are usually found among pond vegetation from Maine to Florida, and west to the Great Lakes and the Mississippi valley. The pumpkin seed feeds on any small animals, such as insects, worms, fish, and crayfish, while the chain pickerel has gained the reputation of eating anything it can overcome. The pickerel has earned this notoriety because of its large mouth, well-armed with sharp teeth — a species the fisherman knows to treat with respect!

The abundance, distribution, and diversity of lake life are influenced by temperature, oxygen, nutrients, and light. Aquatic life is the richest and most abundant in shallow water, where sunlight can reach the bottom. Common producers along the edges of lakes include submerged plants such as the bladderwort *Utricularia* and the green alga *Chara*. These submerged plants in turn serve as cover for many of the scavenger inhabitants of shallow water, including crayfish. The dominant consumer organisms of open lake waters are fish, such as the lake trout, bullhead, pumpkin seed, yellow perch, and chain pickerel.

A wide variety of organisms are capable of living in the diverse lakes across the Northeast. They have adapted to the seasonal changes of spring and fall overturn, as well as temperature and oxygen stratifications. Some are important to us as food sources, while others serve as the basis for intricate aquatic food chains. Perhaps most important for human beings is the way these components intermesh to produce the pristine beauty of our ponds and lakes, which have served for centuries as a source of inspiration and renewal for the human spirit. Thoreau, as usual, sums it up in *Walden*: "I want to go soon and live away by the pond, where I shall hear only the wind whispering among the reeds. It will be success if I shall have left myself behind. But my friends ask what I will do when I get there. Will it not be employment enough to watch the progress of the seasons?"

14.

Freshwater Wetlands

Freshwater wetlands have been viewed traditionally as waste areas, and therefore their natural values have gone largely unrecognized. Biologists estimate that more than one-third of the nation's total original wetland acreage has been obliterated through drainage and filling, and the remaining acres are fast disappearing. Wetlands are interesting for hiking, birding, or botanizing, and serve myriad ecological functions as well. Some wetlands provide important breeding and feeding areas for numerous species of wildlife. Others may supply groundwater for drinking water. All of them play an important role in the hydrologic cycle, the natural cycling of water through the biosphere.

The curious hiker or naturalist may well ask what actually *is* the difference between the various types of wetlands, such as swamps, marshes, bogs, and wet meadows? From an ecological point of view, there are some distinct topographic and biological differences among these four types of wetlands. Biologists define a *marsh* as a wetland habitat dominated by herbaceous plant species, and a *swamp* as a wetland habitat dominated by woody plant species. In both marshes and swamps, groundwater is at or near the surface of the ground for a significant part of the growing season, and thus root systems are annually inundated for long periods. *Bogs* are generally limited to cold and wet climates where wet peat serves as the rooting medium for establishing vegetation. *Sphagnum* moss is usually conspicuous, and standing or slowly running water is near or at the surface during the growing season. A *wet meadow* is often defined as a dense grassland, usually rich in *forbs* (herbs other than grasses), which occurs in a reasonably moist habitat where groundwater is at the surface for a significant part of the growing season.

Several states along the eastern seaboard have passed laws protecting wetlands. In Massachusetts, the Department of Natural Resources filed regulations under the Wetlands Protection Act in November of 1974. By passing such regulations, the Massachusetts legislature accomplished three things: (1) it recognized that wetlands do have value in their natural state, (2) it established a process by which the environmental impact of construction on or near wetlands could be minimized, and (3) it provided local conservation commissions with a mechanism through which to protect neighborhood wetlands. The regulations in Massachusetts cover both freshwater and coastal wetlands, but we will save our comments on the coastal zone for the next chapter.

The Concord River in Massachusetts; in the foreground is a stand of Arrowhead, *Sagittaria latifolia*, a common freshwater marsh plant. / D. O'Reilly

One aspect of the Massachusetts wetlands' law which we as botanists find particularly interesting is the fact that each category of wetland is defined according to the presence of native species of plants. Those plant communities which are legally included as "freshwater wetlands" are: wet meadows, marshes, swamps, and bogs. Plants which are considered indicative of wet meadows include the following: blue flag *(Iris)*, vervain *(Verbena)*, false loosestrife *(Lythrum)*, rushes (family *Juncaceae*), sedges (family *Cyperaceae*), and sensitive fern *(Onoclea sensibilis)*. [While we have used common names of plants and animals whenever possible in this book, we include the scientific names of these plants since the names are included in the law.] Plants considered indicative of marshes include: bur reeds *(Sparganium)*, buttonbush *(Cephalanthus occidentalis)*, cattails *(Typha)*, duck weeds (family *Lemnaceae)*, horsetails *(Equisetum)*, pickerel weeds *(Pontederia)*, water lilies *(Nymphaea)*, and water willow *(Decodon verticillatus)*. Swamps are defined as those areas in which the following are common: alders *(Alnus)*, buttonbush *(Cephalanthus occidentalis)*, hemlock *(Tsuga canadensis)*, highbush blueberry *(Vaccinium corymbosum)*, red maple *(Acer rubrum)*, skunk cabbage *(Symplocarpus foetidus)*, or willow (family *Salicaceae)*. Last, but certainly not least, the term "bog" is used in the law to describe wetlands which are dominated by any combination of the following plant species: sphagnum moss *(Sphagnum)*, black spruce *(Picea mariana)*, bog cotton *(Eriophorum)*, larch *(Larix laricina)*, laurel or sheep kill *(Kalmia angustifolia)*, pitcher plants *(Sarracenia purpurea)*, and sweet gale *(Myrica gale)*.

Let us now look more closely at two representative wetlands in western Massachusetts: Hawley Bog and Lawrence Swamp. Hawley Bog, situated in Hawley, Massachusetts, was recently acquired by Five Colleges, Incorporated, for the purpose of preserving the bog for future study by scientists from Amherst, Hampshire, Mount Holyoke, and Smith Colleges, and the University of Massachusetts. Situated in a depression in the Berkshire Hills, Hawley Bog represents the remains of a lake formed about 14,000 years ago at the end of the last glacial stage in this region. The only vestige of the lake is the "eye" of the bog, which is the open patch of water in its center. As we proceed from the road toward the "eye," we pass first through a climax forest of hemlock, maple, birch, and beech trees. These species now occupy

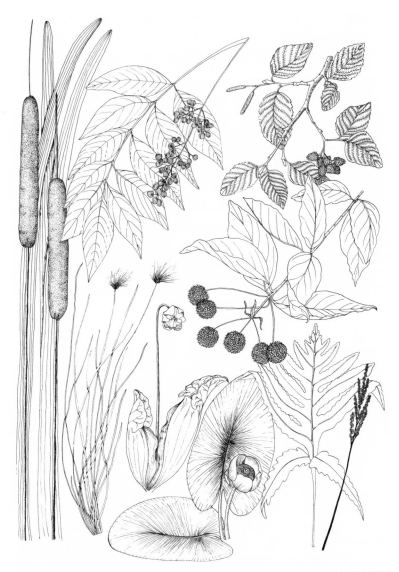

Plant species representative of freshwater wetlands include (starting on the left and moving clockwise around the page): cattail, *Typha latifolia*; poison sumac, *Rhus vernix*; speckled alder, *Alnus rugosa*; buttonbush, *Cephalanthus occidentalis*; sensitive fern, *Onoclea sensibilis*; yellow spatterdock, *Nuphar variegatum*; pitcher plant, *Sarracenia purpurea*; cotton grass, *Eriophorum* species.

what was the shallowest part of the old lake. Moving with increasing difficulty toward the center, we see that the floral composition changes from one group of tree species to another, to shrubs, and finally to typical bog herbage. When we return to the dry outer edge we must remember that this area too has passed in time through the stages we see as we walk toward the center.

If a linear transect (a straight line passing through several zones of vegetation) is run from ten feet out in the open water back 100 feet toward the woods, six different plant zones are encountered. In the open water are floating pads of the yellow water lily. Moving landward, there is a zone of sweet gale, then a zone of cotton grass. The next zone is dominated by sphagnum and carnivorous pitcher plants — these insectivorous species digest small insects trapped within their green, cup-shaped leaves. Another zone is dominated by sheep kill, while the shrub zone closest to the forest edge consists predominantly of Laborador tea plants. Thus, bog succession allows us to witness, by moving horizontally in space, what has happened vertically in time: an excellent example of ecological succession.

Lawrence Swamp is a true "swamp" by definition, and is one of many conservation areas within the town boundaries of Amherst, Massachusetts. Herbaceous plants along the edge of Lawrence Swamp include: royal fern, sensitive fern, Virginia creeper, and square-stemmed monkey flower. The latter plant, as its name suggests, has a square-ridged stem and violet lobe-lipped flowers which resemble a face. Woody vegetation along the water's edge includes swamp rose, alder, red maple, slippery elm, arrowwood viburnum, and buttonbush. The common name for buttonbush comes from the shape of the flower cluster; in July the tips of the branches appear to be decked with white buttons, and later in August or September the buttons assume a yellow-green color as the seed clusters mature. Out in the water are hummocks of various species of sedges, which support steeplebush and meadowsweet.

Quantitative analyses of vegetation in Lawrence Swamp were conducted along four transects in the winter of 1975-1976 by Kathy Torrey, a Hampshire College student. The most important species in the swamp were red maple, hemlock, white pine, muscle tree (also called blue beech), pin oak, gray birch, and alder. During the course of the sampling a few specimens of poison sumac, a close relative of staghorn sumac and poison ivy, were encountered. (Poison sumac has pinnately compound leaves with five or more leaflets and produces grayish-white fruit; it is truly a plant with which all swamp explorers should be familiar!)

Torrey's studies suggest that future succession in Lawrence Swamp depends to a great extent on the activities of humans as well as beavers. If beavers continue to use the swamp, a younger stage in succession, with abundant hornbeam, alder, and dogwood, will probably develop. If, on the other hand, the artesian aquifer (or groundwater supply) underlying the swamp is over-pumped, or if the beavers are trapped or emigrate on their own, the area may become drier and the vegetation may succeed to a white pine, hemlock, and hardwood forest.

The town of Amherst has been most fortunate to have a diligent Conservation Commission whose members have had the foresight to purchase and protect numerous wetlands as conservation areas, thus preserving these areas for all time. We know that many conservation commissions throughout the northeastern United States have been just as active as the one in Amherst and hope that other conservation organizations will follow suit, stepping up activities for the acquisition and preservation of valuable tracts of wetland. It is certainly obvious that people's sense of land values has changed dramatically over the last hundred years — who knows what the true value of tracts of natural wetland may be one hundred years from now?

15.

The Coastal Zone

Life along the edge of the sea is exposed to a group of stresses which result in complex communities that develop in zones parallel to the ocean beach. This can be readily observed, for example, on Plum Island, an accessible and frequently visited strip of land which extends for about nine miles along the western edge of Massachusetts Bay east of Newburyport, Massachusetts. The island was "discovered" by Samuel de Champlain in 1606 — the Indians had undoubtedly known of it earlier — and mapped ten years later by Captain John Smith of Plimoth Plantation fame. A series of fires, excessive grazing, and badly managed agriculture over the years destroyed much of the original pine-oak forest and only on the southern portion of the island, now a part of the Parker River National Wildlife Refuge, has the vegetation begun to return to its natural condition.

If we cross the island, walking from the wet sand at the edge of the Bay to the marsh which borders Broad Sound, the estuary of the Parker River (the Refuge management has arranged a series of boardwalks which allow us to do this relatively dryshod and with minimal damage to plant and animal life), we encounter first a zone of beach, frequently littered with shells, eel grass, and various seaweeds which have been dislodged from their beds offshore. Here, no higher plants are able to survive the wave action and the heavy concentration of salt spray which is produced by the surf and carried inland by onshore winds.

A line of dunes fronts the Bay some yards west. On the crests of these are stands of American beachgrass, beach pea, seaside dusty miller, and scattered individuals of sea rocket and seaside goldenrod. As their very names suggest, these plants are particularly adapted to life on coastal dunes; they are tolerant of salt spray and grow well in response to partial burial by shifting sands. Some, like American beach grass, thrive where sands are active and decline as, through the binding action of the grass itself, the dunes are stabilized. As we progress inland, the dunes continue, interrupted where waves have broken through the frontal series and overwashed the vegetation to create bare expanses of sand or piles of wrack and debris brought in from the outer beach. Windborne salt spray diminishes with distance from the Bay and shrubs become the dominant plants — first beach heather, with wooly gray leaves, and then poison ivy, elsewhere a pest but here a useful sand-binder. Bayberry, beach plum, and several species of wild rose replace the herbs and grasses of the outer dunes, growing taller as salt spray is further excluded, until finally, in the most sheltered portion of the island, a forest community occurs.

Clumps of American beach grass are beginning to stabilize a group of sand dunes at West Gloucester. / AMC Photo Collection

On Plum Island, which has been formed through wave action working over a series of moraines (or ridges of glacial debris), the forest is best developed in several kettleholes, depressions formed as icebergs lodged and melted when the last glaciation ended. Black oak, red maple, and sour gum are common in these woodlands, which, unlike the communities of the seaward dunes, possess well-defined secondary strata of shrubs and herbs below the canopy. The trail guide provided by the Refuge staff for Hellcat Swamp, the largest of these forests, lists blueberry, greenbriar, grape, chokecherry, blackberry, pin cherry, woodbine, raspberry, honeysuckle, and bayberry as common shrubs and vines which provide both food and shelter for the 268 species of birds that regularly occur there.

Westward from the forest, tidal fluctuations play a major role in shaping marsh communities. On Plum Island, a dike has been constructed to halt the flow of brackish water from Broad Sound, establishing a freshwater marsh dominated by cattails and giant reeds. This area was created to provide a favorable habitat for Canada geese, ducks, herons, and shorebirds; unfortunately the value of the marsh for waterfowl is diminished by the spread of purple loosestrife, an introduced species which tends to supplant more useful native vegetation. On the Sound side of the dike, a large salt marsh community exists, extending both north and south of the dike to verge on the road which runs the length of the Refuge to the high moraine at the southern tip.

Plants occurring in salt marshes must be able to withstand some degree of inundation; not all are, however, equally tolerant of salinity. Some, like eel grass and widgeon grass, grow submerged in tidal creeks and pools, continuously exposed to salt concentrations which may be as high as those in the open sea. In areas which are flooded daily, saltwater cordgrass is prevalent, not only on Plum Island but in salt marshes all along the eastern seaboard. In regions flooded only by high equinoctial tides, salt meadow grass, a relative of the cordgrass, forms dense stands which in many marshes, including a stretch above the Refuge boundary on Plum Island, are regularly mowed for hay. Black grass, a rush which is grasslike in general appearance, occurs at the higher elevations in the marsh, while scattered throughout on mud flats are saltworts. Brilliantly tinted in the autumn, these plants concentrate salt in their succulent plant bodies. The delicately branched sea lavender can also be found here. The tidal marshes are the most productive areas within the coastal zone, providing organic matter which forms the basis of marine food webs and acting as sheltered spawning grounds for many species which spend the greater portions of their lives in the open sea.

American beach grass has reached maturity on the crest of these dunes, which face the sea at Plum Island. / D. Hoyt

Plum Island is only one of a group of offshore bars and islands which extend south along the entire eastern coast of North America. Each has a more or less similar zonation from ocean beach to tidal marsh; each can be viewed as a complex of community types put under stress in varying degrees by the marine environment. These islands are, of course, highly prized for recreation and many have been included within national seashores and other state or federal preserves. Paradoxically, the privately owned lands adjacent to these protected areas are most likely to attract exploitative, uncontrolled development and many coastal areas have already been severely damaged by excessive construction and overuse. One need only drive on Plum Island north of the Refuge through the summer colony which lines the road to the Merrimack River to see a prime example of this.

For New Englanders living near the sea, the great blizzard of 1978 provided a vivid demonstration of the destruction which can result from such unwise management. Arriving from the southwest, the storm on February 6 produced winds of hurricane force along the shore. Unusually high tides, in conjunction with the strong winds, broke through natural outer dunes, sea cliffs, and manmade barriers, causing millions of dollars in damage to public property in Maine, New Hampshire, and Massachusetts. Destruction of private property behind the damaged dunes and seawalls was even greater. Nonetheless, by late spring some homeowners were busily rebuilding, setting in motion a sequence of events that can only lead to similar disasters in the future.

A station wagon lies buried in sand and stone in Scituate's Mann Hill Beach area after the blizzard and storms in 1978. / W. M. Brett, courtesy of the Boston Globe

Solutions to the problems arising from the human urge to dwell in coastal areas fall into two broad categories which are structural or non-structural in approach. Structural methods, including dams, seawalls, and other barriers, are designed to protect inhabitants by lessening damage from wind and wave activity. Non-structural methods attempt to keep people away from high-risk areas, maintaining the edge of the sea for its own unique values, including its use as a buffer between the extremes of the marine environment and areas suitable for occupation.

Many problems arise from the use of structural means alone. These devices may be relatively ineffective under extreme storm conditions; in the blizzard of 1978, for instance, some structural devices went unharmed while the buildings behind them were demolished. The use of engineering methods on the beach may have more serious long-term consequences. Seawalls, groins, and jetties, as well as manmade artificial dunes, may, by preventing heavy waves from dispelling a part of their energy on the areas landward from the beach, focus their destructive force on the beach itself, causing more erosion than would have occurred had natural processes been allowed to continue uninterrupted.

A useful alternate approach is one described by Ian McHarg, the well-known town planner and landscape architect, in this book *Design with Nature*. In a chapter entitled "Sea and Survival," he details ways in which man might live on an island similar to Plum Island in a way that would cause a minimum of stress to natural communities. In his analysis, the ocean is viewed as tolerant of recreational uses, although vulnerable to pollution, for which controls are necessary. The beach, being scrubbed out daily by the tides, is also tolerant of recreational use, although intolerant of construction of any sort. The seaward dunes, with their sand-binding, saltspray-resistant herbs and grasses, are, however, extremely intolerant to human activities. No building at all should be permitted on these, and even walking should be confined to boardwalks and bridges. McHarg suggests that limited recreation and construction might be permitted in interdunal troughs where shrubby vegetation has developed. The most suitable area for man, he finds, is that zone of forest which can exist only in the most protected sites. Recognizing the importance of the marshes, he insists that these are intolerant of disturbance, and particularly of filling and dumping. He completes his transect of the island with the observation that the sound or estuary, like the ocean, will sustain intensive use for recreation but must be guarded from pollution.

As biologists we note the virtues of McHarg's arguments. We must also note that the forested areas, which are potentially the most appropriate sites for man, are as well, in many instances, important havens for plant and animal species. The Hellcat Swamp on Plum Island, for example, has been for decades a favorite haunt of birdwatchers who revel in the unusual variety of species which occur there.

Considering the coastal zone in its broadest sense, we cannot help but be impressed with the extraordinary complexity and diversity of its life. Henry Beston, in *The Outermost House*, an account of more than a year spent on a strip of barrier beach on Cape Cod, acknowledges this when he argues for a better perception of animals which ". . . shall not be measured by man. In a world older and more complete than ours they move finished and complete, gifted with extensions of the sense we have lost or never attained, living by voices we shall not hear. They are not brethren, they are not underlings; they are other nations, caught with ourselves in the net of time and life, fellow prisoners of the splendour and travail of the earth."

After standing for more than half a century, the cabin in which Henry Beston wrote *The Outermost House* was washed across Eastham Bar and carried out to sea during a severe northeastern storm. / C. J. Burk

The cabin Beston occupied lasted on Eastham bar for more than half a century, to fall victim to the blizzard of 1978, washed out by storm-driven waves into Nauset Bay. When last seen from the mainland, it was being carried through the inlet to the sea, its fate another demonstration that man's tenure on these offshore bars and islands can be brief at best. By their nature, these features are transitory in themselves; appreciating their uniqueness may be the highest use we can make of them.

16.

Above Timberline

In July of 1858, Henry David Thoreau and several companions spent a week or so exploring the White Mountains, visiting such familiar sites as the summit of Mount Washington and Tuckerman Ravine. The expedition was in part marred by a fire accidentally set in a patch of krummholz, the stunted type of forest typical of alpine zones, by a Mr. Wentworth, who had been hired to maintain camp. The blaze, recorded in Thoreau's journal entry for July 8, began "spreading off with great violence and crackling over the mountain, and making us jump for our baggage. . . it spread particularly fast in the procumbent creeping spruce, scarcely a foot deep, and made a few acres of deer's horns, thus leaving our mark on the mountainside. We thought at first it would run for miles. . . ." After crossing the brow of a ridge east of the travelers, however, the fire burned itself out; nonetheless, the charred "black patch . . . looked like a shadow on the mountain" whenever Thoreau and company looked in that direction for the next few days.

Later in the month, after returning home to Concord, Thoreau systematically organized his observations, setting out the characteristic plants in each zone of vegetation on Mount Washington. The first of these, three-quarters of a mile ascending along the mountain road, was a mixed forest of conifers and hardwoods. At one and three-quarters miles, spruce was dominant with fir and birches. Fir became predominant at three miles up, the limit of well-developed trees, and from the treeline to one mile's distance from the summit was a zone of shrubs or berries. Beyond that was a zone of sedge and cinquefoil and, at the very top, a zone marked by clouds and lichens.

Above timberline in the White Mountains of New Hampshire. / D. Hoyt

As we examine Thoreau's record more than a century later, the resemblance of what he saw to what we might note ourselves in the area is striking. His listing of plants observed might be extracted from the narrative to form a preliminary checklist for *Mountain Flowers of New England*. Ecologists have refined the definitions of zones on the mountain, but all agree on the extent to which vegetational zones change at higher elevations. Furthermore, acts of carelessness such as the runaway fire set by Mr. Wentworth continue to degrade the landscape.

In this chapter, we will consider the ecology of mountains above timberline. We will refer to a number of important plants of alpine habitats, and suggest that readers unfamiliar with the appearance of these might refer to *Mountain Flowers of New England* or other illustrated field guides. (We will consistently use the common names cited in *Mountain Flowers of New England;* since a plant may have more than one common name, its common name may vary from text to text.)

Thoreau was, of course, not the first to describe zonation in mountainous areas. Indeed, he had read the writings of the German Baron Alexander von Humboldt who climbed nearly 20,000 feet to the summit of Mount Chimborazo in Ecuador in 1802. Humboldt, one of the best-known natural scientists of his age, correlated the decrease in temperature at higher elevations with the drastic changes in the vegetation, theorizing that the ascent of a mountain of that height would be equivalent in a biological sense to a journey from the tropics to the poles, with life forms on the summit equivalent to those of arctic tundra.

More recent workers have found it useful, however, to emphasize the differences between alpine and arctic tundra habitats, despite the severe conditions and the many plant species common to both. The term *alpine* is applied to low vegetation above the timberline on mountains. *Tundra,* which means "marshy plain" in Russian, refers to the treeless, flat landscape above the Arctic Circle. In the Arctic, seasonal variations tend to be more extreme and daily variations less extreme than in alpine zones. During the long arctic winter, for example, there are months when the sun does not rise above the horizon; during winters in alpine zones below the Arctic Circle, there is a regular alternation of light and dark within each 24 hour period. In winter, the arctic tundra is generally harsher. In summer, because of nightly heat losses coupled with high winds and intense radiation by day, alpine areas are more severe. Approximately 450 species of flowering plants occur worldwide in either alpine or arctic habitats or both. Within at least some of those species found in both environments, genetically distinct ecological races exist, adapted to differing conditions.

In coastal areas, as we pointed out in our last chapter, zones of vegetation occur in gradients ranging from high to low exposure to windborne salt spray and submergence by the tides. In mountains above timberline, partially as a result of increasing elevation and different degrees of slope and exposure, plant communities sort themselves out into mosaics related to two quite different gradients: (1) increasing winter snow cover and later snow melt, and (2) increasing humidity resulting from fog, soil moisture, and decreased sunlight, which on north-facing slopes retards water loss. A series of extremely useful studies was published on this subject in the early 1960's by L. C. Bliss, who identified nine distinct vegetation types above timberline in the Presidential Range.

Boott Spur from the Tuckerman Ravine Trail. / D. Hoyt

In the most exposed, windswept sites, including the tops of Mount Franklin and Mount Pleasant, a *"Diapensia"* community occurs. *Diapensia* is one of a group known as "cushion plants"; Thoreau noted "it grows in close, firm and dense rounded tufts, just like a moss but harder, between the rocks" on Mount Washington. Much of the ground in this community is bare; characteristic plants in addition to *Diapensia* are highland rush and bearberry willow, forming low but extensive mats.

As snow cover increases, *Diapensia* becomes less abundant, while clumps of highland rush become more prominent; with mountain cranberry, bog bilberry, and three-toothed cinquefoil they form a "dwarf shrub heath-rush" community. Throughout the Presidential Range, this is the most abundant community type, dominating such well-known tracts as the Alpine Garden, Bigelow Lawn, and Boott Spur.

Where snow cover is even greater, highland rush is re-placed by shrubs: mountain cranberry and bog bilberry, crow-berry, Labrador tea, and, in dense mats up to a foot in height, the low, sweet blueberry. This "dwarf shrub heath" community is best seen around the Lakes of the Clouds, on Mount Monroe, and in the Alpine Garden near Tuckerman Ravine. Finally, in areas of greatest snow accumulation and latest melt, often near clumps of *krummholz*, a very diverse "snowbank" community is dominated by dwarf blueberry, tufts of crinkled hairgrass, and a number of herbs more common to the spruce-fir forest at lower elevations. Thoreau observed the snowbank community at Tuckerman Ravine the day of the fire, writing "Here were the phenomena of winter and earliest spring, contrasted with summer. On the edge of and beneath the overarching snow, many plants were just pushing up as in our spring. The great plaited elliptical buds of the hellebore had just pushed up there, even under the edge of the snow, and also bluets." (Thoreau's "hellebore", known to science as *Veratrum viride*, is called "Indian Poke" in *Mountain Flowers of New England*.) Other species commonly found in the snowbank community are Canada mayflower, bunchberry, goldthread, and starflower.

Alpine Flowers. / AMC Research Department Collection

Within the second major gradient, increasing atmospheric moisture, the dwarf shrub heath community is also found, occurring on the clearest, driest sites. With increasing fog at higher elevations, this is replaced by dwarf shrub heath-rush, and then, particularly on north and west slopes around 5,600 feet, by a "sedge-rush-dwarf shrub heath" community. Here highland rush, mountain cranberry, and three-toothed cinquefoil are still abundant, joined by mountain sandwort and Bigelow sedge.

The sedge, a superficially grasslike plant named for Jacob Bigelow, physician, botanist, and early explorer of the White Mountains, has a particularly high photosynthetic efficiency. As a result, it is more capable of making food at low light intensities than other plants. As fog becomes even more prevalent, this sedge increases, displacing the highland rush to form a "sedge-dwarf shrub heath" community. Finally, on the most humid sites on the upper slopes of the higher peaks, Bigelow sedge alone is dominant in a "sedge meadow" community with various mosses and, along trails, the mountain sandwort. Sedge meadow is most extensive at elevations of 5,800 to 6,200 feet on the north and west slopes of Mount Washington. Very wet sites with abundant peat support an "alpine bog" community, also characterized by Bigelow sedge, with sphagnum and other mosses.

Finally, on streambanks is the ninth community type, the "streamside" community with tea-leaved willow in clumps up to two feet in height, bearberry willow in mats no higher than two inches tall, and a very diverse group of herbs.

Most of us who have hiked or climbed in the mountains have heard at some time the statement that alpine areas contain fragile habitats which are easily disrupted or destroyed. The same statement can also be made regarding habitats in the coastal zone, and indeed we concluded our last chapter with that point. However, while human populations have lived, however unwisely, in close association with the seashore, they have tended to avoid mountains above timberline, particularly in Europe and in North America. The Sherpas and Tibetans are among the few year-round occupants of alpine sites. With the increased interest in hiking, camping, and other forms of outdoor recreation, modern man has begun to invade, at least on a seasonal basis, these areas to an extent that in some locations poses a threat to their existence.

a. **Labrador Tea**
   Ledum groenlandicum

b. **Rhodora**
   Rhododendron
   canadense

c. **Bog-Rosemary**
   Andromeda
   glaucophylla

d. **Leather-leaf**
   Chamaedaphne calyculata,
   var. *angustifolia*
   leaf var. *latifolia*

e. **Dwarf Bilberry**
   Vaccinium
   caespitosum

f. **Velvet-leaf Blueberry**
   Vaccinium
   myrtilloides

Courtesy of the AMC Field Guide to Mountain Flowers of New England

A skeptic may ask, nonetheless, how alpine communities could possibly be fragile. To exist there at all, organisms must be adapted to some of the most severe weather conditions in the world. On Mount Washington, for instance, winds of greater than hurricane force have occurred during every month of the year since records have been kept. Clouds cover the area much of the time between sunrise and sunset, often creating dense fog. During July, the warmest month, the mean temperature is about 48 degrees Fahrenheit; during January, the coldest month, mean temperature is slightly less than 6 degrees. (For comparison, July at Berlin, New Hampshire, to the north but at a lower elevation, averages 18 degrees warmer and January 9 degrees warmer than on Mount Washington.) Therefore, plants and animals which live in alpine areas either are extremely resistant to stress, can avoid these conditions by dormancy, or, in the case of birds and larger mammals, move to lower elevations when the season is most rigorous. However, despite the innate toughness of alpine vegetation, at least two interrelated factors cause the system to be very finely balanced. These are low productivity and slow rates of succession.

The total amount of energy which can be fixed by plants in alpine zones is relatively low. A square meter of alpine vegetation can produce no more than four grams of dry matter per day over a two month growing season, while in tropical habitats an equal area may yield four times this amount daily through the entire year.

Succession, as noted previously, is both the process by which soil is formed and a major mechanism by which plant communities recover from disturbance. On Mount Washington, Bigelow sedge invades the mossy covering of rocks as humus in the soil increases. In areas where a lack of snow cover has caused frost action, *Diapensia* and other plants stabilize the surface, to be replaced in time by the dwarf shrub heath-rush community. These are instances of succession, but, because of the low productivity and concommitant slow plant growth, the process takes place very slowly. Some alpine areas have already been damaged by fire or overgrazing; others (and most of us could suggest at least one sad example) are threatened by overuse by man. Any recovery here will proceed at a rate far less rapid than at lower elevations. Without protection, recovery may not occur at all.

In the Presidential Range there is a total of less than eight square miles of alpine habitat. This is the largest alpine area in the eastern United States. Its preservation, given the vastness of the nation, is a goal so obviously worthy that we hope we need not defend it.

17.

Water Pollution

Ecologists who study water pollution and its effects are fond of quoting Samuel Taylor Coleridge, who counted in the city of Cologne, ". . . two and seventy stenches/ all well defined, and several stinks!", and asked "Ye Nymphs that reign o'er sewers and sinks. . . what power divine/ Shall henceforth wash the river Rhine?"

The problem is indeed an old one and still far from being solved abroad or locally. Pollutants can be classified in a variety of categories, and it is helpful for our purposes to distinguish two broad classes, those which act as poisons to living things and those which stimulate plant growth to such an extent that fresh-water ecosystems are unbalanced and *cultural eutrophication* occurs. In addition, there are pollutants which are nuisances, often quite unpleasant ones, but are otherwise for the most part harmless.

Poisonous pollutants may be chemical, physiological, or biological in nature. Proteins, fats, soaps, and carbohydrates are found in domestic sewage; in wastes from creameries, canneries, and slaughterhouses; and as byproducts of numerous industrial processes. Released into streams and lakes, they provide food sources on which disease-producing organisms can thrive and, in their decomposition, consume oxygen needed for the survival of fish, insect larvae, crayfish, mollusks, and the other consumer members of the food chains of healthy bodies of fresh water. As these substances decompose, they also release foul smells which may do little direct damage to anything but the aesthetic quality of life for those persons who live, work, or attempt to find recreation in their immediate vicinity. Oils, tars, acids from the wastes of mines or factories, alkalies from chemical, textile, or tanning industries, and various salts released by manufacturing or road treatment may directly poison living organisms or may change the stream, pond, or lake environment in such a way that organisms may no longer survive in it.

We have studied a freshwater marsh over a period of three years since a single winter oil spill. The site, a part of Arcadia Wildlife Sanctuary in Northampton, Massachusetts, is important as a breeding place for fish and waterfowl, as a resting area for migratory ducks and geese, and, as managed by the Massachusetts Audubon Society, as an educational resource for surrounding communities. The oil, an estimated thousand gallons accidentally released from a large state institution upstream, coated the stems and roots of plants in the zone of emergent vegetation and left a residue where it came in contact with the silty bottom. The following summer, the staff of the sanctuary's natural history day camp reported that few aquatic snails or insect larvae were available for study. Birdwatchers complained that nesting wood ducks, black ducks, and herons were scarcer than before. We found that plant life where the oil had lodged was less abundant than in previous years and that several kinds of plants, particularly annuals which develop from seeds each year, had disappeared completely.

The next year the situation, compounded in part by an unusual midsummer flood, seemed even worse. Perennial reeds and rushes, which yearly develop aboveground parts from fleshy, food-storing roots and rhizomes, had fared better than the annuals the first year. Apparently their energy sources were exhausted in the effort, however, and they too decreased in abundance. With a loss of producer organisms in the marsh, less food was available for other portions of the food web. Moreover, the bottom sediments, no longer tied down and protected by vegetation, were increasingly exposed to erosion. By the third year, the situation had markedly improved, although the vegetation still had not returned to its state before the spill. We see at present two general lessons to be learned from this experience: first, pollution effects are complicated and long-lasting; second, these effects are, in time, reversible when the sources of pollution are removed.

An oil spill occured in the Mill River in Northampton, Massachusetts, during the winter of 1971-1972. Oil was carried into a marsh at Arcadia Wildlife Sanctuary where vegetation was severely damaged. The photograph above was taken during August, 1971, prior to the spill; note the abundant grasses in the foreground. The photograph below was taken during August, 1972, at the same site. Note that vegetation is poorly developed; the grasses are virtually absent and much of the surface of the marsh is bare. / C. J. Burk

Coloring materials, turbidity and suspended particles, changes in temperature, radioactivity, and foam are physical pollutants which can act, either directly or indirectly, to harm freshwater organisms. As the depth to which light can penetrate is reduced by dyes or by tiny particles of silt and clay (these latter, as we have pointed out previously, are too often the results of soil erosion), the amount of food produced by aquatic plants is also lessened. Silt may accumulate on vegetation or in the gills of fish and invertebrates; coal dust may actually weigh down and crush the eggs of fish and other animals. Increased water temperatures may encourage excessive growth of sewage fungi and water plants and, because warmer waters hold less oxygen, may have adverse effects on fish.

Biological pollutants include, besides the pathogenic organisms associated with raw sewage, the bacterium *Closteridium botulinum*, which thrives in waters of low oxygen content and gives off a toxin deadly to geese, ducks, swans, and other waterfowl.

The second major class of pollutants is those which are not in themselves directly harmful to aquatic life. Every gardener knows that the commercial fertilizers he purchases have a code of three numbers, 23-19-17, for example, or 25-15-25, which refer to the percentages of nitrogen, phosphorus, and potassium available (remembered in sequence by the mnemonic *Never Pester Kids*, based on the symbols of the elements, N, P, and K respectively). These are the elements used as nutrients by plants; compounds of nitrogren and phosphorus are particularly apt to cause problems in fresh water.

We have elsewhere discussed the process of aging in lakes, the sequence during which these substances become more abundant, productivity of organic materials increases, and succession occurs, eventually changing open water to dry land. A number of man's activities have speeded up this natural process to the extent that within the last few years certain lakes and streams have become no longer suitable for water supplies or for recreation.

Phosphates and nitrates, the critical compounds of these essential elements, are released in sewage, in runoff from feedlots, and from commercial fertilizers which are too often applied in excess and leached out in runoff water from fields. The old foaming-style detergents caused a form of physical pollution which, at its worst, resulted in drifts and billows reported to be up to thirty feet in height on the lower Mississippi. The biodegradable types which replaced them decompose in water to release phosphates. Plant growth in many lakes, ponds, and streams is limited by the supply of phosphorus available; when more of this substance is supplied, plants grow with great luxuriance. Waterways may become choked with vegetation; "blooms" of blue-green algae may color the water in late summer, giving off toxins which may poison fish and emitting foul odors. Decaying plant materials use up oxygen in the same way as decomposing proteins, fats, and carbohydrates from other sources, resulting in the same injuries to various forms of aquatic life.

For several years, in addition to work at the oil-damaged marsh at Arcadia Wildlife Sanctuary, we have studied Lake Warner, a highly eutrophied impoundment in North Hadley, Massachusetts. Lake Warner was created in the seventeenth century by damming the Mill River, a tributary of the Connecticut which flows through North Amherst and North Hadley. As late as the early 1950's, the lake was widely used for fishing, boating, swimming, and other forms of recreation, a portion of its lower perimeter lined with handsome New England frame houses. Then, as the population of Amherst swelled with the growth of the University of Massachusetts, raw sewage entered the Mill River upstream in increasing amounts. The lake is centered, moreover, in rich agricultural lands supporting lush fields of tobacco, squashes, asparagus, potatoes, cucumbers, and other crops, all of which receive heavy applications of commercial fertilizers in early summer. By the mid-1960's, the upper portions of the lake were a thick and teeming broth of blue-green algae topped with a sheet of duckweeds, tiny floating aquatic plants which thrive on heavy nutrient concentrations. By late summer, the lower portions become almost as densely clotted as the upper reaches.

Although swimming is impossible (or at least extremely unwise) and the water unfit for drinking, the lake is in some ways a botanist's delight. Marsh plants in the upper reaches grow luxuriantly, although the duckweeds tend to overrun the floating leaves of spatterdocks and water lilies. Shrubs along the bank are burdened with pale orange strands of dodder, a parasitic relation of the common morning glory which can stimulate its host to absorb increased amounts of phosphate from the soil. One of the several duckweed species, the watermeal, *Wolffia columbiana,* is extremely interesting. The smallest of the flowering plants and no bigger than a pinhead when mature, watermeal was first discovered in Massachusetts in the 1930's. Now it frequently reaches pest densities in lakes and ponds which receive heavy sewage inputs. Scientists studying the North Hadley lake bed estimate that if all pollution sources were eliminated, the lake still holds a sufficient accumulation of phosphorus to support this growth of vegetation for five years or more.

The surface of this lake is almost totally covered with vegetation. Blue-green algae and duckweeds thrive in the polluted waters, which receive heavy inputs of nutrients from sewage and from fertilizers carried by runoff from surrounding fields./C. J. Burk

Lake Warner is an extreme example, yet it illustrates a pattern of change which is occurring with increasing frequency in many portions of the country. Studies have shown that, as with those sorts of pollution which have directly toxic effects, cultural eutrophication is reversible with time. If nutrient inputs are removed, if sewage is properly treated, and if phosphates from detergents are kept out of aquatic ecosystems, those systems will retain the capacity to recover. How long they will retain this capacity with continued neglect is a question to which one hopes that future generations will not have had to find, through our negligence, the answer.

18.

Noise and Sound

"One man's music is another man's noise," the old saying goes. Like many shopworn aphorisms, it contains an important element of truth — noise is a subjective phenomenon. A simple definition incorporating this subjective element states that noise is unwanted sound.

Two individuals can hear the same sound simultaneously, yet only one may experience it as noise. Similarly, an individual can hear the same sound in two different locations, but only in one place will he characterize it as "noise." Compare, for example, what your reactions are to hearing motors in city streets and in the deep woods.

To gain a more scientific understanding of noise and to discover why certain kinds of noise annoy us, we must first examine sound. Sound is a form of energy which moves through air in a wave motion at 1,100 feet/second. We can describe sound in terms of *pitch* and *amplitude*. The former is a measure of the number of waves reaching the listener in a given period — the greater the number, the higher the pitch. Amplitude describes the strength of these waves — i.e., the height of their crests — and is closely related to loudness. Whether or not a given sound is considered to be "noise" depends on pitch, duration, locale, manner of repetition, and most important — loudness.

Scientists express loudness in terms of watts, a well-known measure of energy. However, if this were common practice the range of audible sound would be very large — from one billionth of a watt for a soft whisper to ten million watts for a rocket engine.

In order to make this range more manageable, the relationship between sounds of different loudness is expressed using logarithms. The result is the *decibel* scale. The $\log_{10}N$ (logarithm of a number to the common base, 10) is the number of times 10 is multiplied by itself to give that number. For example, $\log_{10}100 = 2$, since $10^2 = 100$.

The decibel level of a given sound is computed according to the following formula.

$$\text{Decibel Number} = 10 \times \log_{10} \frac{\text{Power in watts of a given sound}}{\text{Power in watts of a barely audible sound.}}$$

An ordinary conversation is one million times louder than the least audible sound, so in this case the fraction in the formula reduces to one million. The logarithm of one million to the base 10 is 6, and this multiplied by 10 gives us a 60 decibel reading for this sound.

Humans find that higher pitched sounds seem louder, and thus a subjective element that compensates for this phenomenon is incorporated into most decibel scales. Scales zeroing in on different annoyance factors (planes, music, trucks) have been established by individuals active in noise abatement efforts.

Exposure to noise has a number of undesirable effects. To begin with, it interferes with modes of communication: e.g., conversation, music, danger warnings. Noise also disrupts the silence that people at times purposely seek. For example, those who visit the winter woods for their silence may be disturbed by passing snowmobiles.

Doctors tell us that protracted exposure to loud noise (80 decibels and above) causes hearing loss. Eventually, the exposed individual loses his ability to discern quiet sounds. Decibel levels of 125 and above can deafen completely.

Only in technological societies do women hear better than men. This is attributed to the differentiation of labor and technology which places men in closer contact with noisy machines than women.

In addition to loss of hearing, exposure to noise can have harmful effects on general physical health and psychological well-being. Noise creates stress in individuals, which in turn

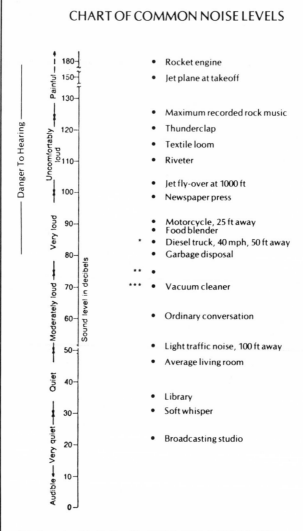

CHART OF COMMON NOISE LEVELS

Sound level in decibels		
180		• Rocket engine
150		• Jet plane at takeoff
130		
		• Maximum recorded rock music
120		• Thunderclap
		• Textile loom
110		• Riveter
		• Jet fly-over at 1000 ft
100		• Newspaper press
90		• Motorcycle, 25 ft away
		• Food blender
	*	• Diesel truck, 40 mph, 50 ft away
80		• Garbage disposal
	**	•
70	***	• Vacuum cleaner
60		• Ordinary conversation
50		• Light traffic noise, 100 ft away
		• Average living room
40		
30		• Library
		• Soft whisper
20		• Broadcasting studio
10		
0		

Danger To Hearing

Painful
Uncomfortably loud
Very loud
Moderately loud
Quiet
Very quiet
Audible

*Maximum legal decibel level for ORVs in NH, July 1973-July 1978

**Maximum legal decibel level for ORVs in NH, July 1978-July 1982

***Maximum legal decibel level for ORVs from 50 ft., in NH, after July 1, 1983

Chart adapted from Turt, Turt and Wittes' *Ecology, Pollution, Environment*, 1972.

Noise level chart

induces anxiety, hormonal and heartbeat alterations, and blood vessel constriction. Work efficiency is lowered demonstrably.

A number of methods exist to reduce the harmful effects of noise. First, we can act on the source. This usually involves limiting by law the hours of operation or substituting a quieter machine to perform a given task.

A second strategy is to interrupt the path of the sound. Sound moves through the air in waves. Materials like wool which vibrate only slightly when struck by sound waves are ideal noise barriers. They are said to possess acoustical qualities. When built directly into the source, these materials are known as mufflers. Most engines used in transport utilize such noise control devices.

Noise can also be deflected away from the receiver rather than absorbed by acoustical materials. Modern highway planners utilize this noise abatement strategy in designing highways which must pass through residential areas.

A final control method entails protecting the receiver. Sophisticated ear plugs and muffs have been developed for industrial workers. Victims of everyday noises cup their hands over their ears or hop the next train to a quieter spot.

Prior to the establishment of the federal Environmental Protection Agency (EPA) in 1970, noise abatement was solely the concern of state and local governments. Ordinances were established limiting the levels and hours of use of noise emission sources. Legally, these regulations were hard to enforce because of their subjective nature. Technological advances in monitoring devices made possible the enactment of quantitative ordinances, but purchasing test equipment and training the personnel to use it was costly. In many areas enforcement of laws was lax in the absence of a specific complaint. Unfortunately, people in our mobile, machine-oriented society was coming to regard loud noise as inevitable.

Today the EPA and a number of enlightened municipal and state governments are working to change this notion. A much publicized EPA report issued in 1972 stated bluntly, "For many city residents noise may be the single most pervasive environmental pollutant." That same year Congress passed the Noise Control Act, giving the EPA the authority to establish federal noise emission standards for products distributed in interstate commerce and to pass this information on to the public. Major noise sources identified in the act include transportation vehicles, motors, appliances, and construction equipment.

The EPA was also given the responsibility for coordinating research into the physiological and psychological effects of noise on humans and animals, monitoring technology, and noise emission control. In addition, the Agency was to provide technical assistance to state and local governments in purchasing monitoring equipment, training monitors, and drawing up noise control legislation. Finally, the EPA was charged with establishing aircraft noise emission standards (with Federal Aviation Agency approval) and with acting as the public's source of information on noise pollution.

In the years since passage of the Noise Control Act the EPA has moved toward fulfilling these responsibilities. Noise emission standards for a number of products including trucks, buses, and motorcycles have been issued. More important, approval of the act has created new resolve in industry and state and local government to deal effectively with noise pollution. Manufacturers of non-regulated machines are voluntarily seeking methods to reduce noise output. Cities and states are increasing the funds allocated to train noise control personnel and to purchase monitoring equipment. Since 1978 financial and technical assistance for state and local noise control programs have been available from the EPA through the Quiet Communities Act. A recently announced "National Strategy for Noise Control" has identified 75 decibels as the maximum 24-hour noise level citizens can be exposed to, and looked to 55 decibels as the future maximum.

State governments are beginning to take action to assure that trail bikes and snowmobiles are not permitted to spoil the backcountry experiences of hikers and backpackers. Forest Service spokesman Ken Sutherland noted that the state of New Hampshire has established strict noise emission limits for Off Road Vehicles (ORV's), and that these regulations can be enforced on National Forest land. Snowmobile and trial bike groups have pledged their support of enforcement efforts. Recently the Forest Service permanently closed the White Mountain National Forest to ORV's except in designated areas.

We are learning that exposure to noise is not the price we

must pay for the comforts of technological society. Without undue optimism it is safe to predict that in future decades our cities and backcountry will be quieter and therefore healthier places to live in, visit, and enjoy.

19.

Solid Waste

The slogan "Carry In — Carry Out" was popularized a few years ago by the Appalachian Mountain Club and the United States Forest Service in an effort to keep the White Mountains and other backcountry areas from acquiring the littered appearance of too many American roads. Today the battle against litter is not over in the backcountry, and in some cities the trash problem is escalating rapidly.

The terms "trash" and "litter" describe objects which are no longer useful or desirable, hence discarded or abandoned and most likely to become a nuisance to others. Household trash includes combustible materials such as paper, wood, or leaves, and non-combustible items such as bottles, cans, and appliances. In this time of rising gas and oil prices, more and more wood and paper are being burned in fireplaces and wood-burning stoves, while leaves often go into backyard compost piles. But, the problem of disposing of the household non-combustibles, cars, agricultural wastes, and industrial wastes (commonly lumped together as "solid waste") is not as simple.

The proliferation of such convenience packaging as throwaway bottles and cans is a relatively recent phenomenon which, according to the soft drink industry, has been dictated by consumer demand. The increased affluence of Americans seemed at one time to dictate a preference for paying for packages, rather than for taking the trouble to return them. Now, at least in some parts of the country, consumers have expressed interest in resurrecting the returnable bottle. Regional recycling centers have sprung up across the landscape. In the early 1970's junior and senior high school students in Northampton, Massachusetts, for example, set up a recycling project operating one Saturday of each month. By mid-1972 the city of Northampton had endorsed the student project and had erected a permanent facility for the regional recycling center at the city yard. Today the recycling center collects, on the average, 30 tons of glass a month, along with large quantities of newspapers and magazines.

Many new synthetic substances used in bottles and packaging materials (particularly plastics and corrosion-resistant coatings for metals) were developed to be resistant to chemical changes, thus ensuring that they would not deteriorate during their useful lifetimes. However, this resistance persists after the products are discarded. In natural ecosystems bacteria and fungi serve as decomposers, breaking down dead plant and animal remains, and thus recycling nutrients needed by other organisms. Manmade compounds would be biodegradable only if sometime during the course of evolution bacteria acquired the capability to break them down; until just a few years ago, however, bacteria would not have encountered any of these synthetic materials in natural environments. Several American companies have spent considerable sums of money during recent years in research toward the production of plastic beverage bottles which are lightweight, non-breakable, returnable, and have a reasonable shelf life. However, if the consumer should (accidentally, of course) discard these bottles along roadways, they would also be biodegradable. This is a start, but much more research is certainly needed in the area of bacterial degradation for a variety of compounds.

Old cars are a different story. In 1970 over 72,000 cars were abandoned in New York City, with the number increasing each year. This number is not unique to New York; other large cities like Boston, Chicago, Philadelphia, and Detroit have problems with junked cars left along city streets, and the cost of removal falls on the shoulders of the respective city taxpayers. Although cars in rural areas are certainly not as numerous as those in cities and are usually concealed more readily along dirt roads or behind the native flora, they still can be a nuisance. One non-profit group in western Massachusetts has been earning money for its charitable activities by locating abandoned cars in its neighborhood and arranging to have these cars hauled away by a metal salvage dealer. But, only certain parts of these abandoned cars are worth recycling; the rest ends up in dumps.

The problem of disposing of household non-combustibles, cars, and agricultural and industrial wastes is not simple. The Northampton city dump, pictured here, was positioned alongside the Mill River floodplain until it was closed in 1969. When a dump is located on a floodplain, erosion and removal of refuse by flood-waters become dangerous possibilities. / R. Hubley

The problems of disposable beverage containers and abandoned cars are closely correlated with the growth of our highly industrialized society, but surprisingly enough, one of the largest single sources of solid waste in this country is from agriculture. One of the principal problems in managing feedlots, where cattle are fattened for market, is waste disposal. In her second edition of *Biology*, Helena Curtis estimates that each steer excretes 65 pounds of manure per day, a total of 230 million pounds per year for a feedlot with 10,000 head of cattle. Other authors have shown that one hog produces as much waste as two people, while seven to ten chickens produce as much waste as one person. At that rate, a chicken farm with 100,000 birds produces as much waste as a town of 15,000 people. The increasing scarcity of land for waste disposal and the characteristics of manure pose some challenging technical problems for livestock producers and, indirectly, the rest of the population.

Several methods of solid waste disposal are currently utilized in industrialized countries, including dumping, landfilling, incineration, composting, rendering, pyrolysis, and industrial salvage. The simplest and most primitive waste repository is the open dump. Waste is collected, usually compacted, hauled to the dumping site, and spread on the ground. However, the dump is a potential health hazard because of the flies and rats that multiply and spread diseases there. Moreover, rainfall tends to wash out large quantities of potential water pollutants from the dump, including dissolved and suspended matter and pathogenic bacteria.

A healthier method of land disposal is the sanitary landfill, in which every layer of refuse is covered with a layer of soil or gravel. The waste should be well shredded and compacted to allow bacterial decomposition to take place efficiently. Disadvantages of both the sanitary landfill and the open dump are the loss of some non-renewable resources and changes in the topography of the area. Often dumps and landfills are positioned in marshes or along floodplains with little regard for the loss of these productive ecosystems. In an earlier chapter we discussed the value of floodplains in serving as natural flood control areas; a dangerous possibility for erosion and removal of refuse during flooding exists if a dump is located on a floodplain.

Incineration, rendering, and pyrolysis are three methods of waste disposal which involve heating refuse in some manner. Incineration is used in many metropolitan areas and involves burning refuse in a furnace to produce an inert and odorless residue. The incineration equipment can handle an assortment of garbage and rubbish without prior separation and can reduce the volume of solid waste by about 80%. Rendering is an example of recycling animal waste in which fat, bones, and feathers are cooked to produce one substance used in animal feed and another used in making soap. Pyrolysis is the practice of decomposing materials by heating them in the absence of air; the biggest advantage of this method is the closed nature of the system, which prevents discharge of potential air pollutants.

One disposal method for residential solid waste which has been successfully employed in several foreign countries is composting. Organic materials can be broken down by bacteria into a sanitary humus-like product, which can in turn be utilized for agriculture. Three conditions for proper operation of a compost plant are: (a) oxygen, which can be provided by tumbling the compost; (b) moisture, guaranteed through the use of sewage sludge; and (c) absence of non-combustibles such as metals, glassware, and ceramic articles. One of the largest composting plants in the world operates in Rome, with a capability for handling up to 600 tons of refuse each day. So far, large-scale composting efforts in the United States have not been successful.

Considerable thought has gone into development of a plan for the disposal of municipal solid waste in Massachusetts. In the early 1970's Arthur D. Little, Inc. of Cambridge was retained by the Massachusetts Department of Public Works to study the feasibility of various recycling strategies for solid waste and to recommend the best alternative. The study, completed in December of 1973, suggested that Massachusetts adopt an intrastate resource recovery system which would provide a significant energy source for the state.

At the time of this writing, the Massachusetts Bureau of Solid Waste Disposal is occupied with the implementation of a resource recovery system in the greater Lawrence area. The proposed site of the new facility is a parcel of land in North Andover, Massachusetts. The resource recovery plant will serve approximately fifty communities in northeastern Massachusetts and southern New Hampshire, ranging from Rockport, Massachusetts in the east, to Derry, New Hampshire in the north, to Westford, Massachusetts in the west, and Woburn, Massachusetts in the south. The plant will recover energy and materials from an expected 3,000 tons per day of mixed residential, commercial, and non-hazardous industrial waste from the fifty cities and towns. Ferrous metals can be recovered from the process and marketed, while the high-quality residue produced by the facility will be landfilled. Perhaps the most valuable product generated by the plant is enough electric power to displace an estimated 860,000 barrels of fuel oil per year. This facility, scheduled to begin accepting solid waste for processing in the early 1980's, has been designed to meet local, state, and federal regulations — the plant will use an evaporative cooling system with no discharge to the Merrimack River, and will release low sulfur emissions to the atmosphere. The production of energy from trash seems especially worthwhile in a state so dependent upon others for its fuel supplies.

Solid waste is indeed a perplexing problem. One would hope that, through education, people could be trained not to throw cans and bottles along roadways or into waterways and encouraged to recycle as many resources as possible. More research needs to be undertaken to unravel the intricate pathways of microbial breakdown of chemical compounds, and manufacturers should be encouraged to incorporate this knowledge into the synthesis of biodegradable materials. We might seriously consider at some point the abolition of all dumps, ending their threats of air pollution through smoldering dump fires or water pollution through leaching of dissolved and suspended matter. The regional resource recovery system appears to be one promising trend for the future. To be "regional" implies a tremendous amount of cooperation among participating communities, but it would seem that if "garbage" could be transformed into energy, that cooperation (and whatever concessions went along with it) might justify the effort.

20.

Air Pollution

Many of us who are interested in the outdoors tend to associate air pollution with urban environments and ignore it if possible or denounce it emphatically when necessary. (Remember those moments just after that snowmobile, blowing clouds of foul exhaust, roared past you on your last winter snowshoe walk?) However, to assume that most of the time it poses no major threat out in the country is almost certainly wrong. Recent work in New Hampshire within the Hubbard Brook ecosystem, a carefully studied small watershed in the White Mountains, suggests that forest trees have been growing more slowly during the last twenty years. Species studied include sugar maple, yellow birch, beech, spruce, and fir. The blame for the decline has been placed, with good supporting evidence, on drought and air pollution.

The air pollution involved takes a particularly insidious form since its harmful effects may well take place through the medium of acid rain, the product of sulfur dioxide (SO_2) and other oxides which, in contact with water, form acids, including sulfuric acid (H_2SO_4) as secondary pollutants. The source of the sulfur oxides is apparently combustion of fossil fuels in metropolitan areas well to the south of New Hampshire; the gas, ironically, is introduced high into the atmosphere by tall smokestacks which were designed to reduce air pollution in the cities. The same phenomenon has been observed in Scandinavian forests where the sources of the pollutants are factories in England and the Ruhr.

Air pollution is a problem that has been with mankind for

centuries. Medieval cities were notoriously smoky, the problems increasing with the increased use of coal for fuel in the fourteenth century. "Smog" has been the bane of modern city life, producing startling episodes of illness and death during periods of its highest concentration. There are, in fact, two distinctly different kinds of smog; in this chapter we will discuss the materials which comprise these and then examine other, perhaps subtler and more dangerous, forms of air pollution.

For nearly a century open burning at this city dump contributed to air pollution in western Massachusetts. The dump has since been closed; the municipal wastes in the area are now buried in a sanitary landfill operation. / R. Hubley

"London smog" has been the most common form in Europe and in eastern and midwestern North America. Its major components are the sulfur oxides produced by burning fuels. Combined with these are "particulates," chiefly finely divided carbon particles which are the easiest components of smog to measure, since they fall out of the atmosphere as soot. The major effect on humans of London smog is some sort of irritation of the lungs. Under particular atmospheric conditions — stagnant air masses, heavy fogs, temperature inversions which prevent warm polluted air from rising — disasters such as the Donora, Pennsylvania incident, where seventeen people died and nearly half the population became ill during a three-day spell of heavy smog, have taken place. During the great London episode of 1952, to cite another example, 4,000 deaths above what would have been expected occurred during a two- to three-week buildup of air contaminants.

"Los Angeles smog," the second form, results from a series of chemical reactions taking place in the presence of light. The raw materials for this smog are hydrocarbons, derivatives of petroleum products released as combustion products of automobile engines and refineries. These hydrocarbons, more or less harmless in themselves, react with oxygen and oxides of nitrogen in the presence of sunlight to produce a series of secondary pollutants whose most direct effect on man is irritation of the eyes and nose.

Carbon monoxide (CO), one of the components of Los Angeles smog, deserves closer examination as a form of air pollution. Odorless, tasteless, without color, the gas in relatively small concentrations reduces man's sensitivity to environmental stimuli, slowing reactions and contributing to accidents. Carried in the bloodstream, it causes heart damage and lowers energy and endurance. In the city of Los Angeles, death rates in the population show a clear positive relationship to the amount of carbon monoxide in the atmosphere; nationwide, more than 1,000 Americans are killed directly by the gas every year, dying as suicides or victims trapped in areas of lethal concentrations. Although the poisonous nature of carbon monoxide is well known and although man's activities add an estimated 200 million metric tons of it to the atmosphere yearly, little was known until recently about how ecosystems respond to inputs of this gas. In the early 1970's, a series of experiments established that the soil itself acts as a "sink" for carbon monoxide, removing it from the air. These experiments further showed that the living organic components of the soil, most probably several species of bacteria, are chiefly responsible for this activity. A sample of soil from a coastal redwood forest, for example, consisted of about 25% organic matter; this soil was more than six times more effective in taking up carbon monoxide than a sandy soil taken from a weed field and containing less than 2% organic matter. Thus, we see again the interrelationships between the various components of our environment; one of the major consequences of soil erosion is the loss of organic matter and a subsequent decline in soil fertility. Not only is man's food supply threatened by misuse of the soil; also endangered is one mechanism for removing toxic gas from the atmosphere.

The Environmental Protection Agency sets maximum permissible standards for six major air pollutants: sulfur dioxide, suspended particulates, carbon monoxide, hydrocarbons, nitrogen oxide, and the "photochemical oxidants" produced in Los Angeles smog. Recommended action to reduce the levels of these in the atmosphere involves the use of cleaner fuels, limitations on traffic within cities, antipollution devices on automobile engines and elsewhere, and changes in industrial procedures. The world energy situation, the world economy, and the very nature of our society make full implementation of these recommendations difficult, if not impossible, a situation we will consider in later chapters.

Just as erosion is a natural process accelerated by human activities, there are natural air pollutants. The eruption in 1883 of the volcano Krakatoa off the Java coast threw a cloud of fine dust high into the atmosphere; the dust particles circled the globe for several years, resulting in spectacular sunsets as well as colder temperatures, including a season in some parts of New England with frost reported every month of the year. Pollen from wind-pollinated plants is a natural air pollutant which brings much misery to man. Ragweed, the principal agent of hay fever (don't blame the goldenrods, which bloom at the same time and have showy but insect-pollinated flowers!) is particularly abundant in the vicinity of cities and in early stages of old field succession. Critics of some environmental policies have pointed out that New York City spent $500 million in 1971 to control the six air pollutants for which the EPA set standards and only $5,000 that same year for ragweed control, despite the considerable hazards ragweed pollen may present to health. Ragweed can be controlled, these critics stated, in large part by hand-pulling, a task that might have provided wholesome summer occupations for inner city youths.

Nonetheless, the series of dangerous or potentially dangerous substances in the atmosphere continues to increase through man's activities. Lead, used in antiknock compounds in gasoline, may reach unsafe levels in urban areas, posing a particular threat to infants and small children, who may acquire the element as well from other sources. Carbon dioxide, one of the raw materials for photosynthesis, is usually found at concentrations of only .03% in the air. It is one of those gases, however, which trap heat above the surface of the earth contributing to the so-called "greenhouse effect." Increased combustion of fossil fuels, themselves products of photosynthesis in the geologic past, has recently increased carbon dioxide concentrations worldwide. What the long-term effects of this increase will be is a matter of controversy. Some scientists predict a worldwide warming trend, melting the polar ice caps, and a rise in sea levels which would drown our present shore line; others predict increased precipitation which, falling as snow at the poles, might trigger another ice age; still others suggest that the warming effect of the greater carbon dioxide concentrations will be counterbalanced by a decrease in heat, as particulates lessen total radiation to the earth. Whatever happens, it is evident that man is tampering with his environment in ways which may have extraordinary and possibly catastrophic consequences.

At the time we are writing this, a new form of air pollution is causing much concern — pollution of the stratosphere, that portion of the earth's atmosphere at an altitude of from about 15 to 50 kilometers in which temperature increases with increasing height. The oxygen (O_2) we breathe contains two atoms of oxygen per molecule; the critical gas affected in stratospheric pollution is ozone (O_3), with three oxygen atoms. Ozone in even relatively small concentrations in the lower levels of the atmosphere is troublesome. It is the principal irritant to human eyes and mucous membranes in Los Angeles smog, inflicting nose and throat discomfort at levels commonly encountered within cities. Since the 1950's, it has been found to be the cause of a number of serious and widespread diseases which affect crop plants, such as onion blight, grape stipple, and weather fleck.

In the stratosphere, however, ozone plays a role essential for continued life on earth by screening out potentially harmful ultraviolet radiation. Three seemingly unrelated products of modern technology produce as byproducts substances which are capable of destroying the crucial ozone layer: supersonic transports and nuclear weapons, which release nitric oxides, and aerosol cans which release substances known as "halomethanes" that break down in the presence of light to form free chlorine. Both

The pollen of ragweed forms an air pollutant which causes allergic reactions in humans. The flowers of ragweed are nondescript and greenish in color; the pollen is abundantly produced and dispersed by wind. Goldenrod flowers at the same time as ragweed and is often unjustly considered to be the cause of hay fever. The pollen of goldenrods is, however, carried by insects, which are attracted to the showy blossoms. The pollen grains of both ragweed and goldenrod are covered by tiny barbs (inset).

free chlorine and nitric oxides cause ozone to be reconverted to O_2 through an elaborate series of reactions. Various estimates of the damage which potentially might occur on earth if ultraviolet radiation should be increased following disintegration of the ozone layer include an increase in the incidence of skin cancers, premature aging of the skin in certain human populations, and possible deleterious effects on climate, vegetation, and the growth of plankton in the sea.

Environmentalists and some ecologists are occasionally accused of being too alarmist in their prophecies, of crying wolf too often. We feel convinced that an understanding of the ecological principles involved, coupled with wise management, would ultimately lead to solution of most of the problems we have been considering. Soil conservation techniques properly applied will prevent erosion, for example; cultural eutrophication has been shown to be reversible. The scope of the dangers we perceive in air pollution leaves us less sanguine. Too little is understood about too many basic processes; too many seemingly harmless substances turn out to have drastic effects. Raymond Dasmann, a noted wildlife ecologist and conservationist, wrote in 1971, "I fear that we may have to face a truly major air-pollution disaster before we take the action that is needed." Even beyond this, we see the problem of insufficient information on which to base a course of action. It would seem one cannot be too cautious at this time.

21.

Open Space

Air, water, and noise pollution, as well as the proper disposal of solid wastes, should be of concern to all responsible citizens. The maintenance of open space as a resource, however, is a goal which is just as important, although it has received less public attention than the other issues mentioned above. The concept of open space is broad and somewhat diffuse, involving issues which have psychological, social, economic, and ecological components. The physical areas which must be considered vary from vast expanses of wilderness accessible only to the experienced backpacking enthusiast to mere scraps of land within our cities supporting a few green plants and, hence, providing contact with natural phenomena in an environment too frequently manmade, bleak, and sordid.

We cannot, in a book of this length, begin to examine the question of *why* open space is important to us as human beings. We should note that some people assign open landscapes little meaning in their lives and can, on occasion, give eloquent and frequently convincing justifications for these feelings. Others behave as spiritual descendants of the early explorers, trappers, or pioneers, obsessed with the need to seek ever wilder country. We will confine our discussion to some middle ground, considering aspects and examples of open space for those who recognize its importance, even if only intuitively, but who live substantially within a modern, largely urbanized society.

Privately owned areas of open space, vital as they are in the mosaic of sites which comprise the open space resource, are inherently vulnerable to development. The continued open status of these lands is dependent on some sort of official designation, protection, and support. At the federal level, the National Park Service at the beginning of this decade administered over 30 million acres of land, about half within the National Park system and the remainder in a variety of parcels ranging from battlefields and cemeteries through seashores, scenic riverways, trails, and the White House lawn.

View of Mt. Desert Island from Cadillac Mountain. / M. Holland

The first National Park east of the Mississippi River was authorized and established in 1919 on Mount Desert Island in Maine. First settled by the Jesuits, the island, with its magnificent panoramic vistas from Cadillac Mountain, was held successively by the French, the English, and an assortment of private owners. The last of these, a group of summer residents known as "rusticaters," donated much of the present Acadia National Park to the federal government, motivated at least in part by a fear that various enterprises, particularly lumbering, might in time overrun the area.

By way of contrast, Cape Cod National Seashore, the other large tract administered by the National Park Service in New England, was not authorized until August of 1961. The largely unspoiled areas of sandy beach, bearberry heath, pine forest, and diverse wetlands had been maintained up to that time mostly by uncoordinated efforts of individual citizens, towns, private organizations, and the state of Massachusetts. From the outset two classes of land were distinguished on Cape Cod within the authorized seashore boundaries: privately owned properties which would be allowed to remain in private ownership and areas open to the public for its use.

Although Acadia National Park now contains a total of 240 islands within its boundaries, only six of these plus portions of four others (including 28,000 acres on Mount Desert) are owned by the federal government. Whereas the strategy for establishing the Cape Cod National Seashore depended largely on purchases by the federal government, expansion in Acadia may well hinge on an increased use of legal devices known as "conservation easements" granted by adjacent landowners. Conservation easements prevent development while allowing the land to remain in private ownership. Signed, they become a part of the title to the property and remain in force regardless of changes in ownership. Easements are granted to recipient agencies which enforce their guidelines. Through accepting easements, the Acadia Park administration has managed to protect several thousand acres of Maine adjacent to the Park. While the public does not gain access to this property, it does gain assurance that the land will not be treated in ways which might change the basic nature of the Park. The landowners receive certain tax advantages as well as some assurance as to how their land will be used by future generations.

Even within the National Parks themselves, protection of the land can pose difficulties. One of the most ironic problems confronting managers of open space is the result of overuse by an enthusiastic public. Although the original concept of National Parks had only a limited popular appeal, the amount of time spent by Americans in outdoor recreation has increased in recent years so that some parks are in danger of destruction. A growth projection for selected activities has predicted, for example, that in 1976 Americans would go camping on 113 million occasions, compared with 60 million occasions in 1960 and an estimated 235 million in the year 2000. Acadia has suffered, though somewhat less than other parks, from too much use; visitations there increased 15% in the period 1968-1972 alone. (Visitations increased 42% at Everglades and 51% at Big Bend, Texas in that interval). A serious corollary to this overcrowding is the inability of the Park

Service, because of insufficient staff and funding, to open additional camps and trails which might disperse visitors more effectively throughout the region. The same insufficient staff and funding have resulted in failure of the Service to sustain adequate programs for interpretation of the natural history of the Park. As educators, we find this latter particularly distressing, given our belief that only those who can be taught to appreciate and understand the natural environment will be motivated in the long run to work towards its protection.

An overused and damaged campsite in the White Mountain National Forest. / AMC Research Department Collection

It is, of course, notoriously easy to criticize government agencies and notoriously difficult to change their policies, particularly in the rather complicated realm of goals and constituencies in which the Parks exist. It may be useful here to change our focus from a relatively large, federally administered unit to a smaller one under control of a smaller organization. The preserve with which we are both most familiar and which we have mentioned in our chapter dealing with water pollution is Arcadia Wildlife Sanctuary in Northampton, Massachusetts. (The similarity of the names *Acadia* and *Arcadia* has resulted in some confusion, even in printed conservation literature. According to the second edition of *Webster's New International Dictionary, Acadia* derives ultimately from American Indian sources. In the form *Acadie* it was the original French name for Nova Scotia, from which Longfellow's heroine Evangeline was expelled with her fellow Acadians. *Arcadia* derives from the name of a mountainous section of Greece where simple countrymen were reputed to live in rustic happiness. The name was later applied in a romantic manner to "any region or scene of simple pleasure, rustic innocence, and untroubled quiet.")

Untroubled quiet is, however, a particularly inaccurate description of the history of Arcadia Wildlife Sanctuary in Northampton and, had its administration and supporters been blessed with total rustic innocence, it could scarcely have survived. During the earlier years of the century, much of the area was owned by a sportsman who prized the marsh, Arcadia's greatest single asset, as a private hunting ground. He posted "No Trespassing" signs freely on his own and his neighbors' property to avoid sharing the bounty of waterfowl which flocked to dense stands of wild rice in what is still the largest river marsh in the central Connecticut valley. In the later 1930's, after a series of disastrous floods on the Connecticut, the Army Corps of Engineers rediverted back into its former path the Mill River, a tributary of the Connecticut which originally fed the marsh and which had been rerouted upstream during the colonial period in a remarkable if unwise feat of engineering. The Mill River was at that time one of the most highly polluted streams in the state of Massachusetts. As a consequence, the wild rice quickly died and the waterfowl declined in number.

The Massachusetts Audubon Society acquired the land as a memorial gift in 1944. Encroaching on the damaged marsh from the west was a city dump which had been established in the nineteenth century. Through the thirty years following acquisition, the staff of the Sanctuary has done battle with public indifference which threatened to close the facility for a time, with wildfires loosed from the city dump, with proposed highway relocations, with pollution in the Mill River and downstream, and with the sort of malign and mindless vandalism which plagues reserves near highly populated areas.

The prognosis now seems fairly hopeful, nonetheless. Public support, once based on a stalwart core of birdwatchers, has been heightened, in part by the increased awareness of environmental problems in the later 1960's, in part by expanded educational programs and activities. The city dump, after an elaborate series of political moves and countermoves, was closed (the illustrations in our last two chapters, "Solid Waste" and "Air Pollution," show the dump at its worst in the early 1960's). Proposed highway relocations through the Sanctuary have failed as yet to come to fruition, and through a carefully enforced series of controls, the Mill River has been upgraded at least one notch in water quality. Wild rice has returned and with it increased waterfowl. Most promising is a program of land acquisition, which has added a prime stand of floodplain forest in Ned's Ditch (part of an ancient oxbow of the Connecticut), some tracts of upland woods, and — by a type of easement known as a conservation restriction — the raised bed of an old trolley line which crosses the meadows from Northampton to Easthampton, linking the two major portions of the Sanctuary. The restriction, granted to the Massachusetts Audubon Society, allows the Society to police the area, manage vegetation for wildlife cover, and establish hedges for buffers and control. The grantor agrees to retain the area in a natural condition; both grantor and grantee hold access to the area for appropriate recreational, educational, and scientific activities.

Unlike a number of environmental problems, against which the concerned individual citizen can often make little headway, there exist clearcut methods by which each of us can contribute to the maintenance of open space as a resource. Property owners can be effective forces both by their decisions as to what will be done with their land in their lifetimes and as to how the land will be passed on. The practice of soil conservation techniques, prevention of erosion, preservation of habitat for wildlife, and protection of relatively unspoiled tracts are just a few of the sensible ways by which we can endeavor to sustain the quality of the landscapes we control. While not every space can or should be set aside for regulated conservation purposes, we can, by support of proper zoning ordinances, determine the quality of its development. We can align ourselves with one or more organizations which work for the acquisition and protection of open space, and we can, within our local governments, support establishment of green belts, adequate parks, and town or city conservation lands. We possess as individuals peculiar resources suited to this purpose, including our time, even if only used in passing petitions, and that ultimate resource in a democracy, our votes.

22.

Urban Ecology

A major failing of ecology as a science has been its neglect of the urban environment. While there is much to be learned from the study of areas relatively unspoiled by man — the tundra, the tropical rain forests, mountains at high elevations — and while these unspoiled areas, threatened by expanding human populations and technologies, demand attention now, there is also much to be gained by viewing the city, with its interacting populations of man and other organisms and its distinctive set of abiotic or non-living environmental features, as an ecosystem.

We would like in this chapter to consider some of the topics of previous chapters as they relate to the urban environment. Earlier we discussed food webs in nature, the presence in biotic communities of producers which make food, consumers which devour producers or other consumers, and reducers which ultimately convert organic to inorganic matter. Man, a consumer, is the dominant species of the city, yet he depends, despite the recent interest in suburban and porchbox gardening, on food supplies produced for the most part in rural areas often at some distance from his table. Hence, soil erosion in California or poor land management in the midwestern states may directly affect the lives and welfare of apartment dwellers on the eastern seaboard.

Producers, green plants growing outside cultivation, do exist within the city, thriving in parks and vacant lots, on construction sites, along road edges and alleyways, and elsewhere. Sometimes regarded as pests, they nonetheless help to make the environment more "liveable," bringing urban man in contact with more natural systems than those of his creation.

Most wild plants that survive within the city are those which have been associated with civilization for many years. Since civilization in its present westernized form is relatively new on the American continent, it is not surprising that the greater number of wild plants in the city are of foreign origin and have been associated with older cultures for long periods. Hence, the pleasant assortment of wildflowers which invades a vacant lot in Boston may include blue-flowered chicory, St. John's wort, sweet and red clovers, white campion, Queen Anne's lace, and ox-eye daisy, all from Europe. If the area is wet, it may support the showy purple loosestrife, an invader which has created serious problems in some areas of New England and in upstate New York, displacing native vegetation in marshes, wet ditches, and along the edges of ponds. Ragweed, cause of hay fever and a biotic source of air pollution, is a native North American, however, as are goldenrods and the common milkweed.

Even on this barren industrial site in South Boston, colonizing vegetation has begun the process of succession. / D. O'Reilly

City trees must be tolerant of a variety of hostile conditions not generally typical of forest situations. Conifers, including spruces, pines, and junipers, are particularly vulnerable to damage from air pollution and are rarely found colonizing urban sites. Some deciduous trees, including quaking aspen, black and honey locust, lombardy poplar, and Norway and sycamore maples, as well as the tree of heaven, *Ailanthus altissima* (more commonly known to our experience as the "stink tree") frequently become established in waste places and may become sizeable specimens, contributing to a successional development which might lead in time to a bit of urban woodland before new construction or some other form of disturbance results in their demise.

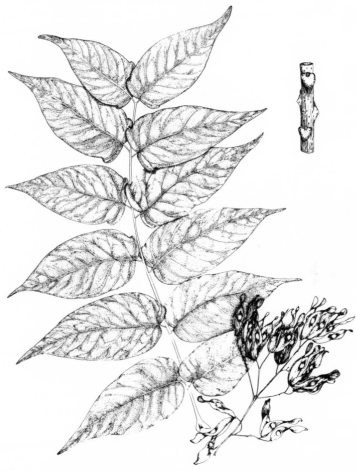

The exotic Tree of Heaven, *Ailanthus altissima*, is tolerant of air pollution and thrives in urban areas where many native trees have failed.

Urban wildlife may be equally varied. Through the late 1950's peregrine falcons could regularly be observed coursing above downtown Manhattan; these elegant predators, now an endangered species (see Chapter 26), bred regularly on the Palisades and nested at least once on the roof of the St. Regis Hotel. John Kiernan's fascinating *A Natural History of New York* discusses a number of forms of life present in that city, including such mammals as the raccoon, the southern flying squirrel, red and gray foxes, and a relatively recent invader from the south, the Virginia opossum, a specimen of which was once reported by *The New York Times* as having "a glorious battle with nine alley cats" on Washington Avenue.

A serious problem of "wildlife" in the city hinges on the prevalence of *feral* cats and dogs, originally pets of man which now roam about with various degrees of freedom. A study of free-ranging dogs in Baltimore, conducted by Alan M. Beck while at Johns Hopkins University, found 450 to 750 of these animals per square mile in the city; they were commonest in areas of high human densities and low per capita incomes. Some of these dogs are pets which are regularly released into the streets for major portions of the day; others have escaped or have been abandoned. Some — an undetermined number — have been born in abandoned buildings, parks, or vacant lots, and have never known a human owner. About half of these dogs roam singly; others occur in packs composed of as many as seventeen individuals.

Large free-roaming dog populations encourage the spread of another urban pest, the common Norway rat, by upsetting garbage cans, providing the rats with an otherwise inaccessible source of food, and by harrassing cats, which might otherwise control the rodents. Public health aspects of the problem also involve increasing incidents of dog bite. (Baltimore city officials report over 6,000 dog bites annually; in more than half of these cases the victims are under age fifteen, with children five through nine at the highest risk. About half the dogs involved are mongrels; German shepherds, collies, boxers, and Dobermans attack humans more frequently than various breeds of hound.) The dogs may act as vectors of human disease as well, spreading measles, mumps, tuberculosis, diphtheria, and scarlet fever. If Baltimore is typical — and Beck believes it is — then free-roaming dogs should be a greater source of public concern than at present. Beck recommends, among other measures, more strictly enforced leash laws, increased control of strays, and a program of general education, particularly at the elementary school level.

Urban animals are of interest for another major reason: fellow residents with man, equally exposed to stress, noise, air pollution, and other problems of the city, they provide us with insights into the effects of urbanization on ourselves.

One of the most vulnerable aspects of the urban ecosystem is its dependence on water. Average household usage is estimated at twenty to eighty gallons per day per individual. Commercial and industrial uses raise the per capita consumption, however, to well over 150 gallons daily, and for this reason virtually all large cities are situated on or near lakes or rivers where a large volume of surface water is available. Nonetheless, in a number of major metropolitan areas, Boston and New York being prime examples, local surface water sources are now insufficient and supplies must be piped in from more distant watersheds. The problems of urban water needs may thus extend to areas far outside the city and acquire extensive political and socioeconomic ramifications, as in the case of the proposed diversion of Connecticut River water from the stream in western Massachusetts to Quabbin Reservoir, source of Boston's public water supply.

Although a number of industries depend upon municipal water systems, there is a marked difference in the quality of water necessary for varying industrial procedures. If water is an important ingredient of the finished product (as in the bakery or dairy industries) or is an agent for cooking, preparing solutions, or washing materials which come into direct personal contact, the quality of water involved must be as high as that which is directly consumed. If, however, dilution and removal of industrial wastes is the major water use and if the water does not come in direct contact with humans, as in paper- and pulp-making, textile manufacturing, steel milling and oil refining, water of lower quality may be adequate.

The city's relationship to water is thus one of user, polluter, and victim of pollution (see Chapter 17); in addition, the use of space in and around the urban area must be coordinated with the place of water in the landscape. Ian McHarg, a noted city planner, has proposed in his extremely useful book, *Design with Nature*, a "hierarchy of urban suitability," a ranking of land types in terms of their potential for development. The most intolerant of extensive human use are fundamentally aquatic ecosystems: surface water, marshes, floodplains, *aquifers* (strata of rock, gravel, or sand which are saturated with ground-water) and *aquifer recharge areas*, the points of interchange between surface waters and aquifers. McHarg suggests that development should be prohibited within the *fifty-year floodplain* of any stream (within a fifty-year floodplain, there is a 2% probability that in any given year the area will be flooded). Suitable floodplain uses include agriculture, recreation, forestry, and various forms of open space. Uses inseparable from floodplain sites — ports and harbors, marinas, water-related industries, and certain water-using industries — may also be permitted. The most appropriate sites for building homes, businesses, and most industries are on flatlands above the fifty-year floodplain. High forest and woodland sites may be used for a number of purposes, but often it is preferable that they, as well as ridges and steep hillsides with slopes in excess of twelve degrees, be preserved as open space. The most significant of McHarg's contributions is his conviction that life within cities — if growth and development are carried out according to what he describes as the "ecological view" — may be both healthy and creative. His arguments are more than convincing. It is time for all those who value the quality of human existence — environmentalists, ecologists, sociologists, city planners, and concerned citizens — to consider the city less a desert, a place to be avoided, and more a challenge for study and improvement.

23.

Energy and Ecosystems

===

Given the current world concern over oil, gasoline, and energy sources in general, we might well turn our attention to the broader aspects of energy, its sources and its behavior in earth's ecosystems.

Early in this book there was a discussion of food webs and the transfer of potentially harmful materials, such as persistent pesticides and radioactivity, from producer organisms, chiefly green plants, through the various levels of animal consumers. Energy also flows through food webs. Unlike chemical components of the web, however, it is not cycled back through the ecosystem but is, instead, finally lost in the form of heat.

The sun, ultimate source of our energy, emits radiation which strikes our planet in a range of wavelengths known as the *solar spectrum*. Those lengths which are actually seen by the human eye, the *visible spectrum*, provide most of the earth's energy when either transformed to heat or fixed in chemical form by the photosynthetic activities of plants.

Only a small fraction, perhaps little more than one percent of the total energy reaching the earth, is captured by photosynthesis; moreover, photosynthetic activities are not evenly distributed upon the globe. In alpine tundra habitats, such as those near the peak of Mount Washington, as little as four grams of dry organic material (somewhat less than the weight of one United States nickel) may be produced daily in a square meter of surface during a yearly growing season of about two months. In a tropical habitat, an equal area may yield almost four times this amount daily throughout the entire year.

Moreover, under a given set of climatic conditions, the amount of energy stored may vary depending on the vegetative cover. Some of the pioneer studies of energy relationships found, for example, that in northern Illinois in the 1920's a cornfield carried on photosynthesis at an efficiency about one-third greater than that of old fields dominated by perennial herbs and grasses in a nearby area. (We will discuss in a later chapter some of the problems of energy in agricultural systems.)

Within the biosphere, energy moves through the food web from plant to herbivore to carnivore with 80% to 90% of the total lost at each transition. Eugene Odum, one of the more important researchers in the field of energy relationships, has illustrated this loss simply yet effectively by means of a pyramid based on 17,850 pounds of alfalfa which supports a calf weighing 2,250 pounds which produces beef to feed a 105-pound boy at the apex. (These calculations are developed in some detail in Odum's third edition of *Fundamentals of Ecology*).

The role of energy in ecosystems has developed into a whole subdivision of ecology, with its own terminologies and methods. In the remainder of this chapter we will concern ourselves with human society and energy — how we use it, where we get it, and what we might do in the future to maintain an adequate supply.

As dieticians and weightwatchers know, a typical human being uses about 3,000 calories of energy derived from food per day. Societies utilize energy, however, in a variety of ways far beyond the nutritional requirements of their individuals — for heat or cooling, for transport, and for the manufacture of products. A large city has been estimated to use fuel at the rate of 4,000 calories per square meter per day, an amount equivalent to the total daily energy input from the sun to an equal area of tropical forest. The source of the city's energy, hence, is not the result of recent photosynthesis but rather of photosynthesis in the past, in the so-called Carboniferous when most of our "fossil fuels," coal oil, and natural gas, were produced.

Most of us are familiar with clubmosses, including the ground cedars and princess pine of northeastern woodlands; horsetails, the scouring rush which invades railroad embankments; and the field horsetail, a common weed of asparagus fields. During the Carboniferous, giant ancestors of these modern forms grew in vast coastal swamps; comparable to though larger than the Everglades or the Great Dismal Swamp, these swamps were located along the southern New England shore and elsewhere on the margins of the continents. One of the more prominent of the swamp plants, *Lepidodendron,* a giant clubmoss, reached heights of more than 100 feet; below it in the understory were smaller herbaceous clubmosses, not unlike those found today. The climate was probably warmer and more humid than at present; the dominant animal forms were amphibians, insects, and primitive reptiles.

The modern clubmoss, *Lycopodium annotinum,* which rarely grows higher than one foot. / D. A. Haskell

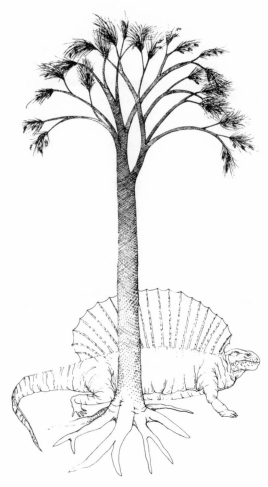

Lepidodendron, an extinct relative of the clubmoss, reached heights of one hundred feet or more and served as the basis for food webs which might have included dinosaurs and other giant reptiles. Our fossil fuels are composed in part of the remains of *Lepidodendron* and its allies.

As the ancient sea level rose and fell, much of this vegetation was covered rapidly by sediments. Food webs were disrupted and the buried plant materials were compressed and converted to the fossil fuel reserves we have today. Some cynics have suggested, only half in jest, that man may have evolved as the earth's mechanism for releasing that energy back into the biosphere.

The technological or industrial revolution of the eighteenth century was marked by man's first large-scale use of these fossil fuels. About 96% of the energy used in the United States today, according to a recent estimate, comes from deposits from the Carboniferous: 43% from petroleum, 33% from natural gas, 20% from coal. Limited though these reserves may be, man has been using them at increasing rates, particularly through the period extending from the end of World War II to the early 1970's. A number of factors were involved in this increased consumption, which occurred both in industry and at the level of individual householders: declining retail fuel prices, increased use of energy-consuming appliances such as air conditioning and automatically defrosted refrigerators and freezers, and, on the part of too many of us, a general carelessness and profligacy.

The Arab oil embargo of the fall of 1973, with attendant price increases, fuel shortages, and rationing, brought the first marked public awareness of energy problems; the bitter winter of 1977 reinforced concerns after a brief and all-too-human inattention. At this point we should all see clearly that political events and natural catastrophes can only heighten a situation inherent in the very nature of our fossil fuel resources: they will only be available until they run out; they were laid down under special conditions and cannot be replaced through modern photosynthesis.

How long our fossil fuels will last has been a point of some contention. Estimates vary in part because of the belief of many workers that use of these materials will be effectively curtailed before the supply is totally exhausted and by predictions that additional deposits may be discovered and exploited. Nonetheless, a conservative recent estimate suggests that at least half our reserves of coal, by far the most abundant fossil fuel, will be consumed in the next 150 years.

It is therefore imperative that alternate energy sources be developed. Those who have considered various alternatives envision in industrialized societies three basic phases of development: (1) a phase to last perhaps twenty years characterized by fuel shortages and planning for the future, (2) a transition phase where coal replaces oil as a fuel and nuclear energy is widely used in the production of electricity, and (3) a phase, perhaps beginning within the next two centuries in the United States, when energy sources are based almost wholly on nuclear fission, nuclear fusion, solar and geothermal power, or a combination of these.

A truism of the environmental movement is the slogan "all power pollutes." Obviously each of these power sources has its problems. We do not have the space or expertise to evaluate them here. Change is, we are convinced, inevitable and the short-term strategy of our generation must involve strong conservation measures. We urge you to cooperate with these.

The ramifications of the energy crisis and its possible solutions are so complicated, controversial, and diverse that additional reading materials are essential for those who would like to understand the situation and its problems further. We suggest the following:

Fundamentals of Ecology, Third Edition, by Eugene Odum (W. B. Saunders Company, 1972).
This solid text, mentioned previously, summarizes much of the scientific research into the nature of energy in ecological systems.

Environment, Power and Society, by Howard T. Odum (Wiley-Interscience, 1971).
Howard T. Odum is Eugene Odum's brother and frequent co-investigator. This book examines the role of energy in human society and contains a number of interesting speculations and ideas.

Energy, the April 19, 1974 issue of the journal *Science* (Vol. 184, No. 4134).
This issue, produced in response to the energy crisis of 1973, contains a number of highly useful articles which discuss the impact of the crisis, problems of economics and policy, and technological aspects regarding oil, coal, gas, nuclear power, the sun, and geothermal energy sources. The issue also contains a useful bibliography. A compendium volume entitled *Energy: Use, Conservation, and Supply* is also available from the American Association for the Advancement of Science, publishers of *Science*.

24.

Human Population
Trends

Recently we have heard some ask why there is concern in the United States today about overpopulation and its associated problems. These people tell us that overpopulation is a problem only in India or in China or perhaps in Mexico, and really is not an issue here. Anyone who has recently climbed Mount Chocorua on a clear summer day or hiked along the Metacomet-Monadnock Trail on a sunny Sunday will realize, however, that the problem of overpopulation is indeed in evidence in the Northeast. Trail overuse and erosion are ramifications both of an increasing human population and the heightened desire of that population to retreat from overcrowded cities and "return to Nature."

In our last chapter we noted that most of our "fossil fuels" were produced in the Carboniferous, a period which is estimated by some scientists to have lasted from 345 to 280 million years before the present. Hominids, or man-like apes, did not evolve until four to five million years ago, and modern man did not appear on the scene until relatively recently (about one to two million years ago). Today the United States, although not the most highly populated country in the world, is the biggest consumer of energy. Indeed, the population of the United States is using a disproportionately large amount of the fossil fuels laid down in vast coastal swamps many millions of years before those early hominids evolved.

Let us take a look at the growth of the human population over the last several thousand years. During the past 15,000 years, anthropologists believe there have been three major increases in human population growth. The first increase is attributed to a tool-making or cultural revolution; the population grew because man learned how to hunt more effectively, and with more food available more of the humans born survived. The second increase is associated with herding and the agricultural revolution; when early man first found about 10,000 years ago that he could domesticate animals and grow more than enough food in one spot by cultivating certain plants, he could also create cities and complex societies. The third increase is commonly attributed to the scientific-industrial revolution. At this time steam, coal, and even water power became widely used, and technology provided a level of prosperity unknown previously. Some people believe that the increase during the past hundred years is especially steep due to the "medical revolution" which has greatly prolonged life expectancies. The development in the 1930's of sulfa drugs and in the 1940's of antibiotics greatly reduced death rates in many countries.

One of the first people to write about overpopulation was the British clergyman Thomas Malthus. In 1798 he published *An Essay on the Principle of Population as it Affects the Future Improvement of Society.* Malthus associated much of the poverty he observed in eighteenth century England with large family size.

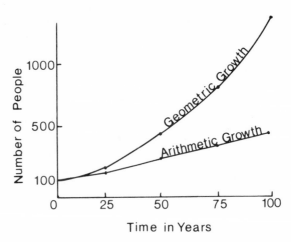

Comparison of geometric and arithmetic growth rates. In 1798, Rev. Thomas Malthus suggested that while the human population increases geometrically, food production increases arithmetically.

One of the main ideas presented in his thesis is that population increases geometrically but food increases arithmetically. Crop productivity can only be increased by small increments through use of fertilizer and more intensive farming, in a manner shown in the series 2, 4, 6, 8, 10, etc. But population growth can increase exponentially, as in the series 2, 4, 8, 16, 32, 64, etc. Eventually growth comes to a halt because of the finite limits of food, space, and pollutants. Malthus went on to state that man could control population by rational thought and through moral restraint, deciding not to have children in hard times. If Malthus had revisited the earth today, after its population has reached four billion people, he might have noted that moral restraint appears to have had little effect.

Statistics from the *Demographic Yearbook* published by the United Nations in 1974 show interesting comparisons between various countries throughout the world. Birth rates were high in the following African countries: Angola (50 births/1,000 people), Liberia (49.8/1,000), and Kenya (47.8/1,000). Birth rates were low in the following European countries: West Germany (10.1/1,000), East Germany (10.4/1,000), and Austria (12.8/1,000).

The rate of increase in population size in Angola is not as high as one might expect because the death rate there (30 deaths/1,000) is among the highest in the world; in fact, life expectancy is just 33.5 years in Angola for both men and women. On the other hand, the death rate in West Germany (11.7 deaths/1,000 people) is higher than the birth rate, resulting in a natural increase of −1.6 per year; life expectancy in West Germany is high (67 years for men and 74 years for women). In our own country, the birth rate in 1976 was 14.2/1,000, the death rate was 8.4/1,000, the rate of natural increase was 5.8/1,000, and life expectancy for men was 67 years and for women 75 years.

When figures on population trends are compared from continent to continent, we find that the two most crowded regions are Europe and Asia, while two of the least densely populated areas are North America and the U.S.S.R. Average population per square kilometer in Europe is 95 people, in Asia 80 people, and in both North America and the U.S.S.R. 11 people. Thus, although some countries in Europe have extremely low birth rates, the continent as a whole is very densely populated.

In the wild, population fluctuations are subject to a natural system of checks and balances. The factors which control population growth are complex and highly interrelated. Certainly food supplies, available space, disease, and predators are all important.

Although North America is one of the least densely populated continents, urban areas tend to be crowded. In this photo, shoppers bustle along a busy street in downtown Boston. / D. O'Reilly

Modern medicine, by reducing the number of human deaths from diseases such as small pox and typhoid, has served to lower the death rate. The resultant lengthening of life expectancies was a most welcome turn of events in humanistic terms, but in many countries it has not been accompanied by a reduction in the birth rate, causing a surge of population in recent years.

Not every government recognizes that overpopulation and overcrowding put increased pressure on natural resources and open space. Much improved communication and education on population-related matters are needed, not just abroad, but in our own country as well. Certainly we can hope that all heads of government will consider population policies which will not only be compatible with the natural, cultural, and social resources within their country, but which will also blend with the goals and ideals of other governments on "spaceship earth." As the former president of the American Association for the Advancement of Science, Athelstane Spilhaus, has said with reference to total population increase, "When we can treat all existing persons as human, it will be time enough to think about having more."

The ramifications of population growth and possible control measures are not only complicated, but are also highly controversial. Thus, we would like to suggest the following for anyone who would like to understand the situation and its problems more fully:

Environmental Biology: The Human Factor by Weldon L. Witters and Patricia Jones-Witters (Kendall/Hunt Publishing Company, 1976).
This paperback places a major emphasis on population, providing introductory information on demography along with discussion of possible future influences of population on food and mineral resources.

The Limits to Growth by Donella H. Meadows, Dennis L. Meadows, Jorgen Randers, and William W. Behrens, III (Universe Books, 1972).
The four authors are members of the Club of Rome, an informal organization whose purpose is to foster understanding of the varied but interdependent components (including population growth) that make up the global system in which we live. Their use of computer simulation techniques is controversial and has stimulated much discussion.

The Population Bomb by Paul R. Ehrlich (Ballantine Books, 1968).
Paul Ehrlich is a biologist from Stanford University whose original research in the area of insect populations led eventually to his grave concern for human population trends. His concern is explained in this paperback. The book is a major document of the "environmental movement" of the 1960's.

Ecology and Field Biology, Second Edition by Robert Leo Smith (Harper and Row, 1974).
This ecology textbook summarizes much of the scientific research on patterns of population distribution in one chapter entitled "Populations: demographic units."

25.

The World Food Problem

Energy, human population growth, and world food supply, the topics of our most recent chapters, are subjects so closely interrelated that they cannot meaningfully be discussed alone. The energy crisis of our decade has at its roots an overdependence on fossil fuels, the products of photosynthesis in the past. The world food problem can only be resolved by the proper management of photosynthesis at present. Without the rapid expansion of human populations following the Industrial Revolution, neither energy nor food on a worldwide basis would be overriding concerns today.

In an earlier chapter, "The Ecological Perspective," we discussed problems related to the role of man in the winter arctic ecosystem, where food chains are compact. Here, during the period of widespread aboveground testing of nuclear devices, radioactive isotopes from fallout were passed rapidly from the lichen producers to reindeer and caribou and then to human beings.

As one moves from the poles towards the equator, food webs in natural ecosystems become increasingly elaborate and difficult to analyze. The webs which support industrialized societies rely on sources of productivity within a complicated network of transportation and are, hence, even more complex. At base, however, these societies depend on the soil resource, which is renewable though frequently misused, and on fresh water and the seas, both of which can be drastically mismanaged. Three basic tasks, therefore, confront those who would attempt to forestall worldwide famine and to assure an adequate food supply for all: (1) to increase the total amount of food grown, (2) to make this food available where and when it is needed, and (3) to limit growth so that human population will not increase to a point where neither of the prior tasks is possible.

A number of methods for increasing world food production have been suggested. Among the more important is the development of the sorts of technologies involved in the "Green Revolution." The Green Revolution resulted from the development and widespread use of high-yielding strains of grain, primarily rice, wheat, and, to a lesser extent, sorghum, in underdeveloped countries. These strains require a high level of nutrient input, usually derived from chemical fertilizers, an abundant water supply, frequently from irrigation, and in many cases the use of pesticides.

There is no question that, properly applied, the techniques which produced the Green Revolution can have an important role in increasing total world food supplies, and indeed Norman Borlaug, an American plant breeder, was awarded the Nobel Peace Prize in 1970 for his part in these advances. However, to those concerned with environmental problems, a mere description of how the revolution is achieved must act as a warning signal; the dangers of misuse of fertilizers and the role of nutrients in eutrophication, the tendency of pesticides to spread through the food web, concentrating at each successive level, and the vulnerability of *monocultures*, areas devoted exclusively to the cultivation of a single species, are too well known to be ignored. In practice, moreover, various social, political, and economic problems have arisen in connection with Green Revolution technologies, including the displacement and impoverishment of small farmers by large landowners who can acquire the needed seed, pesticides, and fertilizers more easily and use them to better advantage.

A second method for increasing food supplies involves developing new varieties of traditional crops and domestic animals. One alternate strategy to the methods of the Green Revolution attempts to develop a number of different strains of crop plants, each bred to achieve maximum success under the peculiar growing conditions of a specific area. Hence, vast acreages devoted to a single variety are supplanted by smaller units supporting a diversity of varieties, each adapted to a particular climate and soil type, each resistant to a particular set of pests and diseases. By this second method, genetic diversity is maintained and the kinds of threats posed — for instance, in the United States in 1970 when almost 15% of the corn crop was lost to corn blight, a disease which attacked a single widely-grown strain — could be avoided.

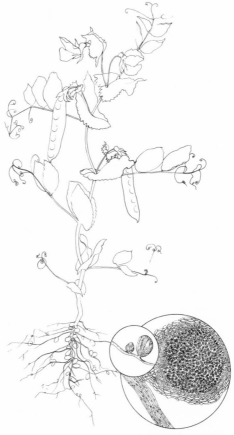

Although nitrogen is the most abundant element in the atmosphere, it cannot be used by higher plants in gaseous form. On the roots of legumes such as the common garden pea in this illustration, nitrogen-fixing bacteria occur in swellings known as nodules (inset). The bacterial cells within the nodules convert nitrogen into forms which legumes can use in growth and other processes. Legumes can thus be raised commercially without nitrogenous fertilizers; when plowed under they add to the available nitrogen content of the soil.

In addition, breeding programs have been established in an attempt to overcome problems caused by excessive use of chemical fertilizers. These involve the development of types of nitrogen-fixing bacteria, such as those commonly found on leguminous plants — alfalfa, beans, and peas — which might form a symbiotic relationship with the roots of grasses, including corn, wheat, and rice, that normally absorb forms of nitrogen directly from the soil.

Human populations can also obtain a more abundant food supply, of course, by lowering their position in the food web. In our discussion of energy and ecosystems we described Odum's famous pyramid in which nearly nine tons of alfalfa support a calf weighing slightly more than one ton which produces beef to feed a 105-pound boy. While this hypothetical boy might balk at a diet of alfalfa, he might be fed quite well at far less energy cost on a diet of suitable vegetable proteins.

Some recent studies have attempted to assess energy costs involved in producing various food items. Most expensive are animal products, particularly those which require extensive transportation (such as products of fishing in distant seas). Least expensive per calorie of food produced are such field-grown crops as corn, soybeans, sorghum, and wheat.

The oceans, which together cover more than two-thirds of the surface of the globe, have been hailed by many as an immense, as yet unharvested, storehouse of food for humans. However, because of the low productivity of the open seas, the immense energy requirements of the fishing industry, and the fact that most marine organisms now consumed by man are highly placed in the extremely long food chains of the ocean, serious workers who have examined the marine resource conclude that its potential as a cure for the world food problem is extremely limited.

What does seem abundantly clear at present is the extent of our reliance on the land and the need for continued and expanded soil conservation practices. To change our emphasis from global to regional concerns, we note the recent decline of farmland in New England. In Connecticut a century ago, 80% of the land was used for agriculture; today the total farmland in that state has diminished to about one-fifth of that amount. Massachusetts, which shows a similar decrease, now imports 85% of its food supply, much of it from California. Here the link between energy and food becomes apparent and one must consider the costs in terms of fossil fuel — ancient photosynthesis — balanced against a more efficient use of photosynthesis now. Would it be advantageous in our overall resource budget to raise foodstuffs in the vicinity of major eastern cities and save the energy involved in long-range transportation?

The open ocean near Beal Island, Maine. / G. Bellerose

To answer this fully, we must examine the abandoned farmlands. A history of Conway, Massachusetts, describes the decline of rural populations from the town by 1867 and earlier as follows: "The population of the town. . . was greatest near the close of the last century. It was at this period that there began a great outward flow of emigration from us to the westward, which has not ceased to the present time. . . this great emigration is not perhaps to be regretted. We may wish, however, that it had not been accelerated and indeed necessitated by the improvident husbandry of the first generations of farmers. The soil was thriftlessly drawn from and its riches spent. The steep and fruitful hillsides were plowed and sowed, and suffered to be washed by the rains, often for many successive years, until they would yield no more. The effects of this wretched culture are still too plainly visible." The area described by the historian is now in various stages of secondary growth and was used in an earlier chapter,

"Man and the Process of Succession," to illustrate the return of forests and associated wildlife to New England. We consider this, in the sense of total environmental quality, a form of progress. If so, if soil misuse and erosion were prime causes of the decline of agriculture in New England and if the re-establishment of forest has been beneficial, then what areas might or should be returned to agriculture?

The U. S. Soil Conservation Service has devised a land classification system that uses soil type, slope, drainage, tendency for erosion, and other criteria to establish the optimal use to which any site should be put. The classes vary from flat, well-drained types suitable for agriculture with no special soil conservation techniques (other than good farming practice) to types totally unsuited not only for cultivation but also for grazing and forestry. Much of the land described by the Conway historian should never have been cultivated.

Fortunately, though, well over 50% of the land mass of the United States can support some form of agriculture; obviously it is this portion which might be brought back into use. Moreover, as the loss of farmland has accelerated, the land given up has gone not to forest but to commercial, institutional, and industrial purposes. Each of us can point to examples of prime farmland which lie beneath the asphalt of a parking lot. It is development, and too often unplanned development, that has in recent decades removed the land from cultivation. Here, constructive changes in policies and attitudes regarding zoning and taxation can and must be sought.

Any solution to the world food problem, therefore, may be seen to demand different approaches in different regions, with an overall emphasis on planning on a worldwide basis for the most effective use of the land, followed by a careful implementation of these plans. In some areas, the technologies of the Green Revolution may prove effective; in others they clearly will not. In addition, since many climatologists now predict that the bad weather of the early 1970's, with widespread drought and unusually cold winters, may persist into the future, an enlarged storage system for food reserves may also be essential. With these methods, our first two tasks, producing more food and making it available where needed, may almost certainly be accomplished. The third task, control of human population growth, looms even more formidably ahead.

We have not discussed the possibilities — to some the likelihood — of world-wide famine by the end of this century nor, indeed, the extent to which malnutrition and actual starvation are problems now. We are not unaware of these issues and urge our readers to examine these matters further. We have found the following readings useful:

Ecoscience: Population, Resources, Environment by Paul R. and Anne H. Ehrlich and John P. Holdren (W. H. Freeman and Company, 1977).
This extensively updated version of a work which was first published in 1970 contains over 100 pages devoted to food in a chapter entitled "A Hungry World." It also contains material on population, energy, and other environmental issues.

Global Perspectives on Ecology by Thomas C. Emmel (Mayfield Publishing Company, 1977).
The nicely integrated set of readings in this book are particularly useful in their approach to the relationships of population growth and food supply in underdeveloped countries and in the tropics generally.

Diet for a Small Planet by Frances Moore Lappe (Ballantine Books, 1971).
In addition to a discussion of the dietary role of proteins, this easily read book contains a number of interesting recipes for individuals attempting to lower their positions in the food web, presenting some alternatives to beef or alfalfa for Odum's 105-pound boy.

26.

The Problem of Endangered Species

In considering *natural resources,* components of the environment which are of use to man, it is often helpful to distinguish resources which are *renewable* from those which, if lost, are forever gone. The first class might include soil, water in oceans, lakes, and streams, and, broadly considered, vegetation and its associated wildlife. *Non-renewable* resources include wilderness, which once disturbed can never return to a truly wild condition, and individual species of plants and animals, each a product of thousands of years or more of evolutionary development which could never be retraced.

Extinctions involving the death of an entire species have occurred with regularity in the past, and indeed great groups of organisms have undergone the process — the dinosaurs and the giant clubmosses whose remains provide our fossil fuels are good examples. Nonetheless, at present the rate at which species become extinct or begin to approach extinction is increasing.

Over the past few decades, the International Union for the Conservation of Nature and Natural Resources (I.U.C.N.) has compiled two lists, a *Red Data Book* which details the status of animals which are in danger and a second listing of species — the *Black Book* — of those already gone. Of the birds and mammals known to be living in the year 1600, the date from which modern extinctions can be calculated with reliability, approximately one percent are now extinct and one out of every forty species is in danger. Scientists predict that by the end of the next half century, more than 100 bird or mammal species will be gone.

Until recently, endangered plant species have received less attention than endangered animals. Nonetheless, of the approximately 20,000 higher plants known in the continental United States, 1,200 are believed to be threatened at present, 750 are in danger of extinction, and about 100 may already be extinct. Again, the rate of elimination may be speeding up. Between 1800 and 1850, only four plant species were known to become extinct. Between 1851 and 1900, 41 were exterminated, while in the first half of the twentieth century, as many species disappeared as in the entire century before.

What can be done to slow these trends? If one examines the causes of extinction, certain patterns are apparent — and some groups are particularly vulnerable.

In the animal kingdom, simply being proportionally large or predatory or restricted in habitat or geographic distribution may render a species more susceptible than smaller forms placed lower in the food web or more broadly tolerant in habitat and range. About one-fourth of the birds and mammals which have become extinct since 1600 probably did so as a result of natural causes and would have declined whether man was around or not. Overhunting by man played the dominant role, however, in the extinctions of the greater number of mammals and large bird species, followed by human disruptions of habitat and the introduction by humans of predators, competitors, or other harmful organisms. Among the small bird species which are now extinct, direct persecution was less a cause than habitat disruption and the introduction of exotics.

Plants of commercial value, especially those such as cacti, orchids, and carnivorous plants, which are removed from their often limited habitats by well intentioned hobbyists or unscrupulous professional collectors, comprise a class which is particularly threatened. In the plant kingdom generally, however, destruction of habitat by logging, excessive grazing and burning, draining or filling wetlands, and growth of urban areas has contributed to the plight of nine out of ten endangered species.

Animals and plants confined to islands face peculiar dangers because of their limited distributions and, in some cases, reduced competitive abilities compared with widespread, more broadly adapted forms which have evolved in mainland habitats and then been introduced to the islands.

Ginseng, *Panax quinquefolius*, left, is a more generally distributed plant in the Northeast. Nonetheless its populations are endangered because of the commercial value of the roots, which are extensively harvested and shipped to the Orient to be used in medicines.

The Furbish lousewort, *Pedicularis furbishiae*, right, is an endangered plant species which occurs on the banks of the St. John River in northern New Brunswick and Maine. Recently the plant has been a source of controversy as its populations have been threatened by proposed construction of the Dickey-Lincoln Dams.

Given the fact that the causes of man-induced extinctions are generally known, alleviation of these causes — controls on excessive hunting and collecting, protection of critical habitats, and the prohibition of the introduction of potentially disruptive exotic plants and animals — is clearly possible. Hunting restrictions date back in North America to the colonial period, although the first state regulations were passed as late as 1877. Federal laws governing wildlife began to evolve during the last third of the nineteenth century; a number of these have been of major importance.

The Migratory Bird Treaty of 1916, for example, provided protection at the international level for game birds moving south from Canadian breeding grounds, while a 1929 Migratory Bird Conservation Act enhanced the treaty by allowing for the development of refuge areas. (Species which regularly migrate across international boundaries or through international waters are another category of those most vulnerable to extinction or endangerment, since they cannot be protected by enactment of a single set of national regulations. The Migratory Bird Treaty was hence far more effective when Mexico joined the agreement in 1937.) Although many of the earlier laws seemed ultimately intended to provide a continuous stock of fish and game for sport and consumption, emphasis has more recently been extended to preservation of non-game species and important types of habitats.

Here again, the animal kingdom has received more attention than the plants. For instance, at the present time four of the nine northeastern states have no laws whatever providing for rare or endangered plants. A Massachusetts law enacted in 1925 established a fine of up to $50 for picking the mayflower (trailing arbutus) and allows the fine to be doubled if the act occurs secretly, at night, or while the perpetrator is disguised. An additional 1935 Massachusetts law protects wild azaleas, orchids, and the cardinal flower. Rhode Island protects approximately a dozen plants while Vermont and New York protect, at least in theory, a larger number of species.

Potentially more important than these laws, which are rarely if ever enforced, is legislation preserving habitats. The Endangered Species Act of 1973 and its ramifications could alone be dealt with at book length. Among other things, the Act provides protection for endangered species, sets up a "threatened" category for organisms which may become endangered in the near future, and directs all federal departments and agencies to ensure that any actions which they authorize, fund, or carry out do not jeopardize these species or destroy or modify their habitats. Many Northeasterners are well aware of how, in the case of the Furbish lousewort and the proposed damming of the St. John River, the Endangered Species Act assumed even broader significance in the realm of environmental problems and citizen response.

On a smaller scale, actions of various organizations, agencies, and even private individuals, have contributed to the preservation of some species. For instance various groups including the federal government and the National Audubon Society have worked together to increase the population of the whooping crane, a commendable example of a form of altruism which may be a new development in human history. In *A Sand County Almanac,* one of the classics of the conservation movement, Aldo Leopold wrote, "For one species to mourn the death of another is a new thing under the sun. The Cro-Magnon who slew the last mammoth thought only of steaks. The sportsman who shot the last pigeon thought only of his prowess. . . . But we who have lost our pigeons mourn the loss." He was writing, of course, of the passenger pigeon, most abundant bird on the North American continent at the time of the European settlement, the last individual of which died at the Cincinnati Zoo in 1914.

Not everyone would agree with Leopold that modern man should mourn these losses and, indeed, some persons, even scientists, have argued that there are species we would be well off without. After all, as we tried to demonstrate in our three most recent chapters, *Homo sapiens* is a species with a population increasing at uncontrolled rates, facing a limited food supply and soon-to-be exhausted sources of energy. We should look to our own safety surely. Why then concern ourselves with other creatures?

For many of us the issue involves moral problems, the question of whether we as a species, with our own long and complicated history, have a responsibility for the welfare of other species with us on the planet. The recent controversy over grizzly bears in the National Parks has brought this issue into focus for many conservationists.

Among the several practical reasons for concern about endangered species is the fact that some threatened or endangered organisms are of real or potential use. For example, non-human primates, including lemurs, apes, and monkeys, are increasingly important in biomedical research as models for the study of human disease. A number of these species are already in danger of extinction before their full value has been realized. Hence, scientists are urging immediate research on their ecological relationships and reproductive biology as the basis for an active conservation program.

Equally vital is the role of species as a sort of "miner's canary" reflecting the state of our environment. Just as the miner's canary showed a sensitivity to harmful gases before the miner could perceive them, so might changes in the welfare of otherwise insignificant species reflect changes in our own environment which might, in the long run, damage humanity.

Moreover, the actual role of each species in the various ecosystems of the world is not known and probably never will be fully comprehended. A species which may seem insignificant today may, in a time of disturbance — changing climate or conditions — assume unexpected prominence. Given the interrelatedness of nearly every aspect of our environment, it would seem the height of folly to let perish what might have, in the future, contributed to our own survival.

Articles describing individual species and their status are fairly regular features of many publications and are usually well-written, accurate, and worthy of attention. Some of the stories are reassuring; others are depressing; most have a certain fascination. Two good source books are *Wildlife in Danger* by James Fisher, Noel Simon, and Jack Vincent (Viking Press, 1969), which is based on the files of the I.U.C.N., and *Extinction is Forever: The Status of Threatened and Endangered Plants of the Americas* (The New York Botanical Garden, Bronx, 1977).

27.

Conclusion

Perhaps more than other scientists, ecologists tend to draw morals, propose warnings, and suggest modes of action based on their research. Although the intent of this book has been to focus on ecology *per se* — i.e., on the study of the relationships of organisms, including man, to their environment — we found, as we concluded each chapter, that one or more of several basic themes recurred:

(1) the extraordinary, often poorly understood, degree to which organisms are interrelated through food webs, nutrient cycles, and other means as parts of ecosystems;

(2) the potential hazards to all life which occur when these relationships, which have evolved over millions of years, are disrupted when harmful substances — pollutants in the broadest sense of the term — are introduced in the environment;

(3) the varying capacities of different ecosystems for recovery through succession and other processes;

(4) the need for careful management of all resources, including soil, water, the atmosphere, open space, and energy, to provide for the basic necessities of human life — and a subtheme to which ecologists are particularly sensitive, the present lack of much critical information related to these resources;

(5) the need for the preservation of particularly scarce, vulnerable, or irreplaceable ecosystems or individual forms of life; and,

(6) the importance of a knowledgeable citizenry in maintaining a viable environment.

In *A Sand County Almanac,* Aldo Leopold several decades

ago separated humanity into "some who can live with wild things and some who cannot." We assume that most of our readers tend to fall into the first category which, in Leopold's words, sees ". . . a law of diminishing returns in progress."

In recent years, many agencies of government have become far more responsive to the concerns of citizens than in the past, setting up mechanisms for input into agency-sponsored programs. Here for all of us is an excellent opportunity to share concerns, insights, and expertise, not only with friends and colleagues but with legislators and government officials as well.

The solutions to many of today's environmental problems are by no means simple. We are, nonetheless, convinced that creative and responsible answers to at least some of the complex problems this book has raised may be obtained through the involvement of members of the business and industrial communities, economists, sociologists, lawyers, legislators, planners, and scientists, along with an educated public, in the decision-making process.

Leopold suggested that perhaps we can achieve a useful change in our values by "reappraising things unnatural, tame, and confined in terms of things natural, wild, and free." And of course Thoreau, whose thought underlies and permeates so many of our own conceptions of the landscape, affirmed, "In wildness is the preservation of the world."

Index